Fluid Power
Educational
Series

Pneumatic Systems and Circuits
-Advanced Level

Joji Parambath

Pneumatic Systems and Circuits – Advanced Level

Copyright © 2026 Joji Parambath

All rights reserved

No part of this book may be reproduced or transmitted in any form or by any means, electronic or mechanical, including photocopying, recording, or using any information storage and retrieval system, without the written permission of the publisher.

ISBN: 9798652506858

https://jojibooks.com

First Edition: 2020
Revised Edition: 2022
Revised Edition: 2023
Revised Edition: 2026

Disclaimer of Liability

The contents of this textbook have been checked for accuracy. Since deviations cannot be avoided entirely, we cannot guarantee full agreement. Only qualified personnel should be allowed to install and work on pneumatic equipment. Qualified persons are those authorized to commission, ground, and tag circuits, equipment, and systems in accordance with established safety practices and standards.

Dedicated to

my son Siddarth

Table of Contents

Chapter	Description	Page No.
--	Preface	vii
1	Introduction to Multiple-actuator Pneumatic Circuits	1
2	Elimination of Signal Conflicts by Idle-return Rollers	8
3	Cascade Method – Design and Operating Principles	12
4	Shift Register Method – Design and Operating Principles	30
5	Stepper Module – Design and Operating Principles	36
6	Review questions	39
7	Additional exercises for circuit development	39
Appendix 1	Graphic Symbols of Pneumatic Components	42
Appendix 2	Designation of Pneumatic System Components	47
--	Solutions for additional exercises	48

Preface

This textbook explains the method for developing advanced pure-pneumatic circuits involving multiple actuators. The starting point in designing any circuit is to have a thorough understanding of the control task. Work out a circuit using a displacement-step diagram or a notational representation of the control task. Then, examine the developed circuit to improve and simplify it.

The problem of signal conflicts and various methods for eliminating them are explained in detail in this book. The development of multiple-actuator circuits using the cascade method and shift registers is illustrated with numerous examples. Intermediate positions in circuits are also provided wherever possible to ensure easy understanding.

This orderly way of developing control circuits should be applied to more extensive control systems. Neat circuit diagrams can be developed that are easy to read or modify when appropriate guidelines are followed and systematic steps are taken. This orderly way of developing a circuit also imparts a certain degree of reliability in the circuit's operation.

Many other fluid power topics are dealt with in other textbooks by the same author. Please see the details at https://jojibooks.com.

Enjoy reading the book.
Your feedback is most welcome.

JOJI Parambath

About the Author....

Joji Parambath has been a trainer in the fields of Pneumatics, Hydraulics, and PLC for over 25 years. During his career, he trained numerous professionals from the industry, faculty members, and students at engineering institutions.

At present, he is the key trainer at Fluidsys Training Center, Bangalore, India (https://fluidsys.org), which provides training in Pneumatics and Hydraulics. He already has two books on Pneumatics and Hydraulics to his credit. The present series of 32 books is intended to restructure and update the existing books.

The author wishes to thank all trainees for their lively interaction and many useful suggestions during the training programmes that prompted the author to write the present series of books. You may send your feedback to info@fluidsys.org.

10th June 2020

Chapter 1 | Introduction to Multiple-actuator Pneumatic Circuits

The development of single-actuator pure pneumatic circuits is described in the textbooks 'Pneumatic Systems and Circuits - Basic Level (in the SI units)' and 'Industrial Pneumatics - Basic Level (in the English units)' by the same author. As explained, compressed air is delivered to an actuator, such as a cylinder or rotary actuator, via a final control element, such as a directional control valve. A pneumatic controller controls the final control element.

An interesting fact that has emerged in the design and development of control circuits is that double-pilot directional control valves are very convenient for more complex control requirements. That is, with double-pilot directional control valves, continuous output actions can be achieved with momentary input signals. Therefore, double-pilot directional control valves are extensively used in pure pneumatic systems.

A typical pneumatic system uses multiple actuators that, when operated in a specified sequence, perform the desired control task. Multiple-actuator pneumatic circuits also conveniently use double-pilot directional control (DC) valves as final control elements. Fully automatic or semi-automatic operation can be achieved using pneumatic sensors that confirm the positions of pneumatic actuators. A pneumatic controller generates the necessary control signals in response to pushbutton and sensor signals to actuate DC valves.

However, it is important to remember that double-pilot DC valves are susceptible to signal conflicts or signal overlap. Hence, the primary requirement in developing multiple-actuator pneumatic circuits is knowledge of ways to eliminate signal conflicts. This textbook systematically presents multiple-actuator pure pneumatic circuits. Symbols of the most important components are also given in Appendix 1.

Please note that the details of the multiple-actuator electro-pneumatic circuits are presented in the textbook "Electro-Pneumatics and Automation" by the same author.

What is Signal Conflict?
Double-pilot valves are convenient for easily realizing many control functions, as they respond equally well to short pulses and continuous signals representing system variables. A double-pilot valve is also called an impulse valve or a memory valve, as a single impulse is sufficient to switch it.

A major problem with a memory valve is its inability to change its switching position when pilot signals are present at both pilot ports simultaneously. These signals produce equal and opposite forces on the valve spool, and hence the spool tends to remain stationary until one of the signals goes off. This problem, called signal conflict or signal overlap, is a major hurdle in developing workable multiple-actuator pneumatic circuits.

Representation of the Control Task for a Stamping Operation
Many of the initial concepts in developing pneumatic circuits with multiple cylinders are illustrated using a stamping operation. The same concept can also be applied to circuits involving rotary or semi-rotary actuators. The stamping operation uses two pneumatic cylinders A and B. The control task can be expressed in the text form, schematic layout, notational form, and/or displacement-step diagram. These forms are explained in the following sections.

Control Task 1.1 | A Pneumatically-controlled Stamping System

Text Form
A stamping operation uses a job-positioning cylinder A (1.0) and a stamping cylinder B (2.0). Cylinder A (1.0) extends, bringing the job (workpiece) under cylinder B (2.0). Cylinder B, then, extends and stamps the job. Cylinder A can return only after cylinder B has retracted fully. A pneumatic control circuit must be developed to implement the control task.

Schematic Layout
The arrangement and installation of pneumatic cylinders for control task 1.1, which governs the stamping operation, can be illustrated using the positional layout shown in Figure 1.1.

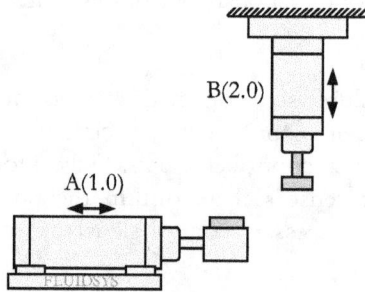

Figure 1.1 | Schematic layout for the stamping operation of control task 1.1

Notational Form
In the notational form, pneumatic cylinders are designated as A, B, etc., and a '+' is added to these designations to represent the forward stroke and a '-' sign to represent the return stroke. For example,
A+ represents the forward stroke of cylinder A
A- represents the return stroke of cylinder A
B+ represents the forward stroke of cylinder B
B- represents the return stroke of cylinder B

For control task 1.1, the sequence of cylinder operations can be represented as: A+ B+ B- **A-**.

Displacement-step Diagram

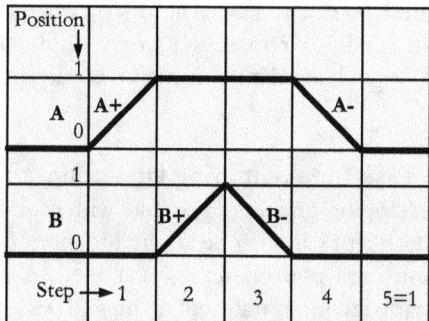

Cylinder Position: 0 – Retracted, 1 – Extended
Figure 1.2 | Displacement-step diagram for control task 1.1

The displacement-step diagram for one cycle of operation, representing the problem given in control task 1.1, is given in Figure 1.2. In this representation, cylinder displacements are plotted according to the required sequence of cylinder operations at equal intervals. A cylinder can move from the retracted position (0) to the extended position (1) and vice versa.

Sequence Control
Figure 1.3 shows the power circuit for the control task represented as A+B+B-A-. It consists of pneumatic cylinders, final control elements, and sensor positions. Two 5/2-way double-pilot valves 1.1 and 2.1 control two cylinders A(1.0) and B(2.0), respectively, through pilot ports 14 and 12 in each valve. For automatic operation in the desired sequence, two sensors per cylinder are used to confirm the end positions of each cylinder's action. The sensor that is actuated in the retracted position of cylinder A is designated as 'A0'. Further, the sensor actuated in cylinder A's extended position is designated 'A1'. Sensor positions for cylinder B (B0 and B1) are also designated similarly.

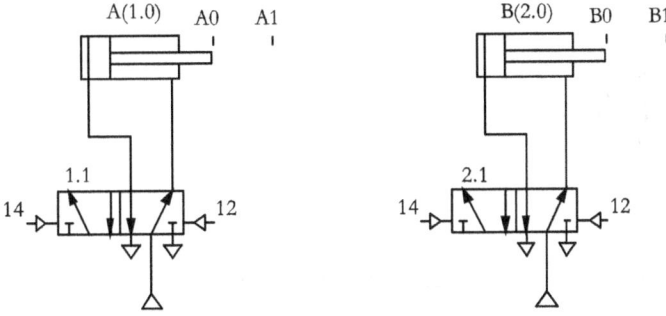

Figure 1.3 | The pneumatic power circuit for control task 1.1

Circuit Design for the Sequence A+ B+ B- A- (Control task 1.1)

Step 1: Representation of the Control Task
The first step in drafting the circuit for the control task is to write down the notations for the desired sequence of cylinder operations and the associated sensor designations, as shown in Figure 1.4.

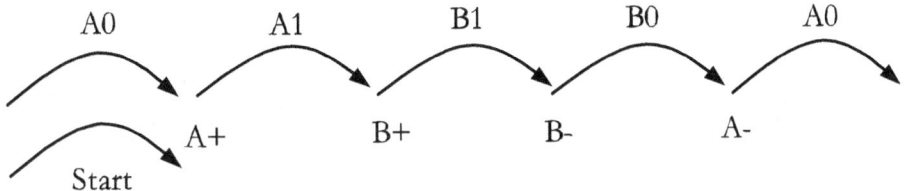

Figure 1.4 | Representation of control task 1.1

From the notational form given in Figure 1.4, it can be seen that:
The A+ action generates sensor signal A1, which is used for the B+ action.
The B+ action generates sensor signal B1, which is used for the B- action.
The B- action generates sensor signal B0, which is used for the A- action.
The A- action generates sensor signal A0, which is used for the A+ action.

It can be observed that the last signal, A0, is used to obtain the first cylinder action, A+. Usually, a 'Start' signal is also required along with the A0 signal to get the A+ action. The next step in designing the circuit diagram is to draw the power circuit and then the control circuit. The developed circuit is then checked for signal conflicts.

Step 2: Draw the Pneumatic Power Circuit

Place the symbols of cylinders A(1.0) and B(2.0), as shown in Figure 1.5. Draw the respective final control elements 1.1 and 2.1, with pneumatic or pilot actuation at both ends (14 and 12) of each valve. Mark the installation positions of sensors A0, A1, B0, and B1.

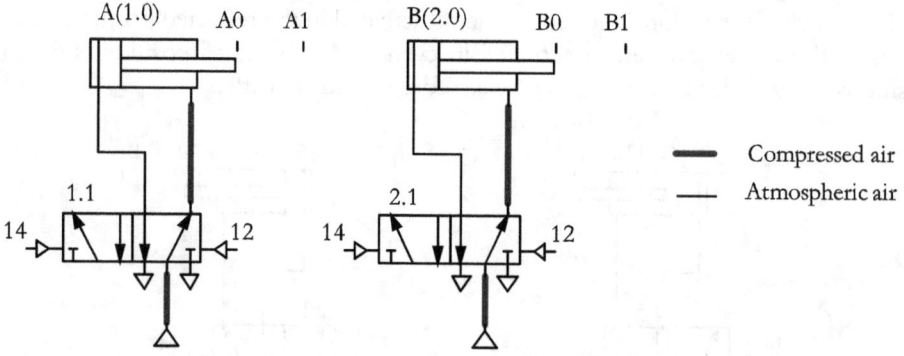

Figure 1.5 | The pneumatic power circuit for control task 1.1

Step 3: Draw the Pneumatic Control Circuit

Incorporate the 'Start' pushbutton and all the sensors in the control part of the circuit as per the required control sequence. Represent sensors A0 and B0 in the actuated condition. The complete pneumatic circuit is given in Figure 1.6.

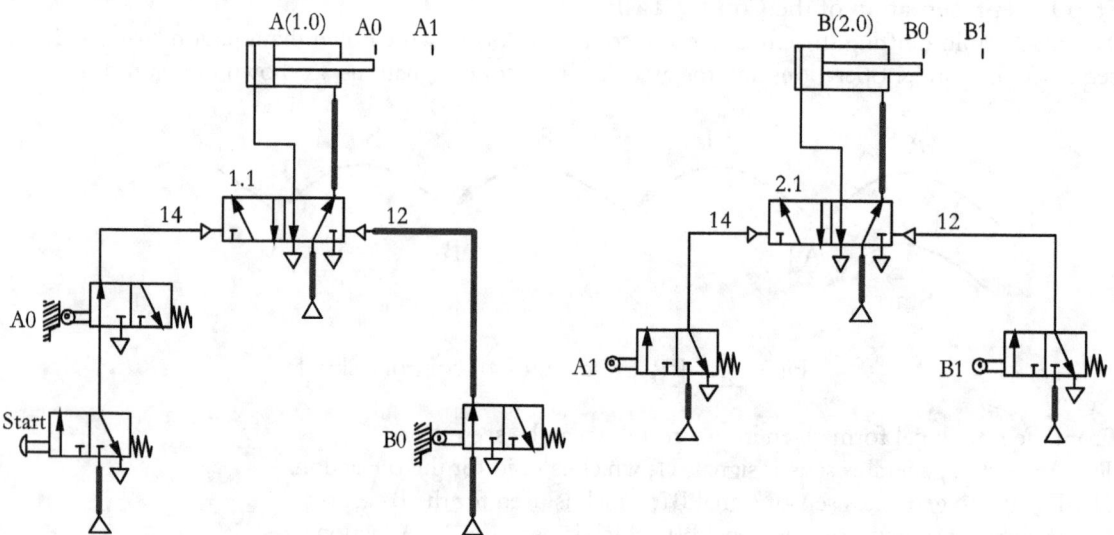

Figure 1.6 | The pneumatic circuit with a control circuit added to the power circuit (Normal position)

Step 4: Analysis of the Control Circuit

The next step is to analyze the circuit for signal conflicts. It can be observed that initially, both cylinders are in the retracted position and hence sensors A0 and B0 generate output signals.

Further, the start pushbutton and sensor A0 are connected in series to implement an AND logic function.

The actuated 'Start' pushbutton generates a signal, and then it logically combines in series with the signal from A0 to appear at port 14 of valve 1.1, as shown in Figure 1.7. Now, check for a signal at the other end (12) of valve 1.1. It can be seen that a signal is already present at port 12 of valve 1.1, resulting in a signal conflict. As a result, valve 1.1 cannot switch over. The signal conflict is due to the persistent signal at port 12 of valve 1.1, resulting from the continued actuation of sensor B0 by cylinder B.

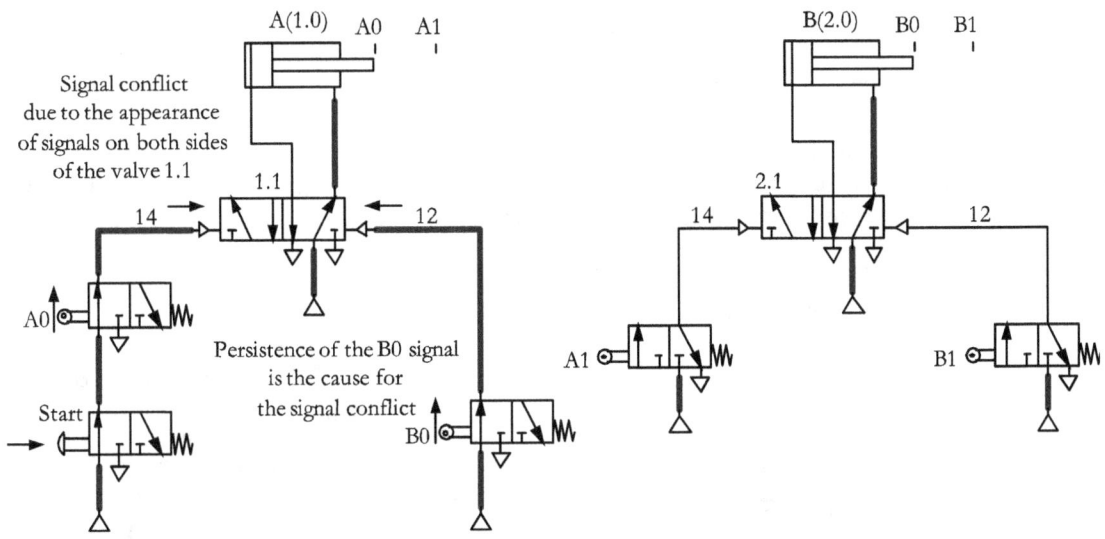

Figure 1.7 | Demonstration of the presence of a signal conflict at valve 1.1

Valve 1.1 can switch over; hence, cylinder A can extend only if the signal at port 12 of the valve is relieved. Assuming sensor B0 is forcibly released, cylinder A can extend. When cylinder A extends (A+), sensor A0 is released.

When cylinder A extends fully, sensor A1 generates an output signal which is then applied to port 14 of valve 2.1. Now, check for a signal at the other end (12) of valve 2.1. It can be seen that no signal is present at port 12 of valve 2.1; hence, there is no signal conflict. As a result, valve 2.1 switches over, and consequently, cylinder B extends automatically (B+).

When cylinder B extends (B+), sensor B0 is released (which was forcefully released earlier).

The circuit position after cylinders A and B are extended (A+ B+) is shown in Figure 1.8.

When cylinder B extends fully, sensor B1 generates an output signal that is applied to port 12 of valve 2.1.

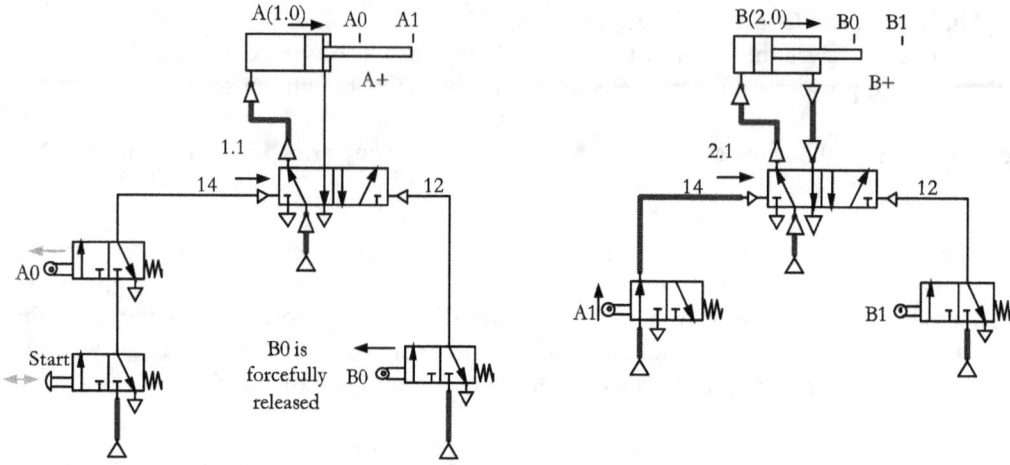

Figure 1.8 | A position of the circuit after A+ and during B+

Now, check for a signal at the other end (14) of valve 2.1. As shown in Figure 1.9, a signal is present at port 14 of valve 2.1, resulting in a signal conflict. As a result, valve 2.1 cannot switch over. The signal conflict is due to the persistent signal at port 14 of valve 2.1, resulting from the continued actuation of sensor A1 by cylinder A.

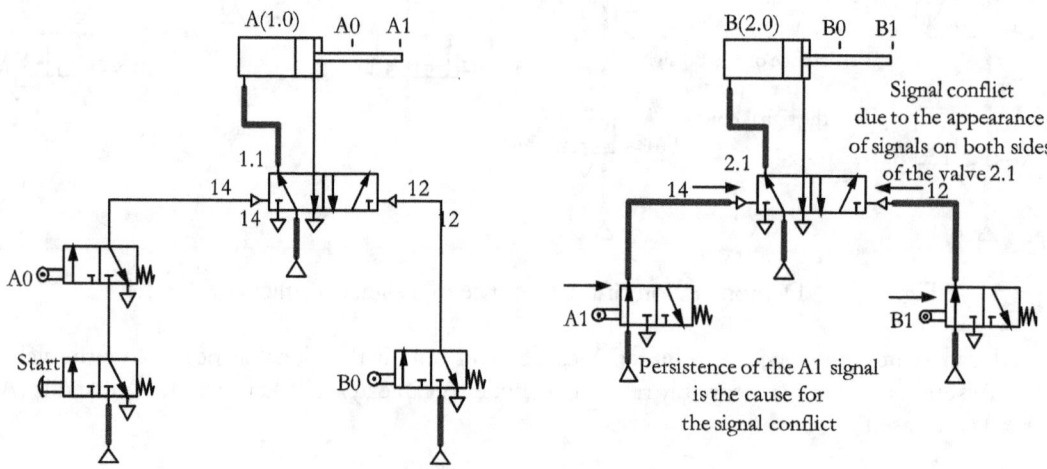

Figure 1.9 | A demonstration of signal conflict at valve 2.1

Valve 2.1 can switch over, and hence cylinder B can retract only if the signal at port 14 is relieved by some means. Assuming that sensor A1 is forcefully released, then cylinder B can retract. When cylinder B retracts (B-), sensor B1 is released.

When cylinder B retracts fully, sensor B0 generates an output signal which is then applied to port 12 of valve 1.1. Now, check for a signal at the other end (14) of valve 1.1. As shown in Figure 1.10, no signal is present at port 14 of valve 1.1; hence, there is no signal conflict. As a result, valve 1.1 switches over, and consequently cylinder A retracts (A-) automatically. When cylinder A retracts (A-), sensor A1 is released.

The circuit position after cylinder B and cylinder A are retracted (B-A-) is shown in Figure 1.10.

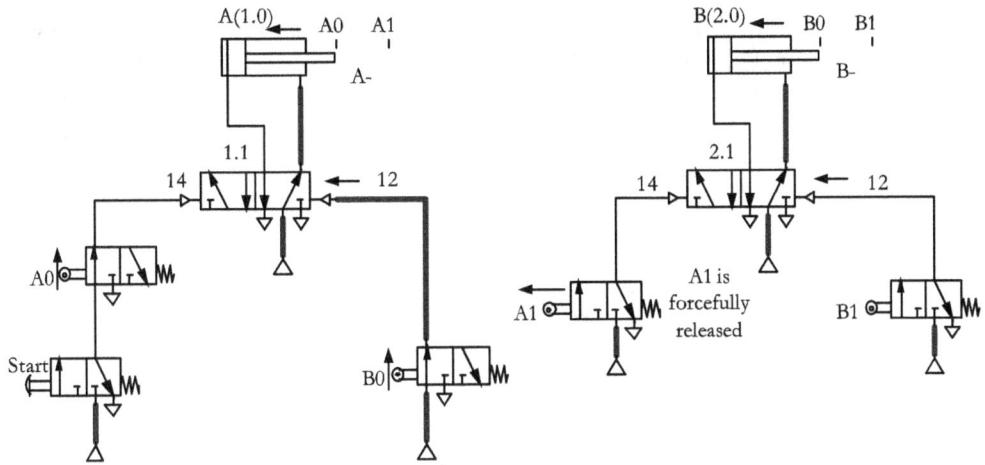

Figure 1.10 | Position when cylinder B and cylinder A have fully retracted (B-A-)

Conclusion

Figures 1.7 to 1.10 illustrate two signal conflicts occurring within a single cycle of operation under the specified sequence for control task 1.1. Consequently, the circuit will not function as intended if these signal conflicts are not eliminated.

The following sections explore ways to overcome signal conflicts.

Elimination of Signal Conflicts

Signal conflicts arise when two or more control signals, without any priority, act on both sides of a double-pilot memory valve and attempt to actuate it.

Many methods have been devised to address signal conflicts in multiple-actuator pneumatic circuits. Some methods rely on controlling the air supply to different sections of the control circuit. The following types of valves/modules may be used to avoid signal conflicts: (1) Idle-return rollers, (2) Reversing valves (memory valves), and (3) Modules as a combination of valves.

Elimination of Signal Conflicts by Reversing Valves

A bi-stable memory valve used in an application to eliminate signal conflicts is called a reversing valve. When using reversing valves to eliminate signal conflicts, the basic idea is to ensure signals are effective only when needed. Two of the possible designs used to remove signal conflicts through reversing valves are:

- Cascade method
- Shift register method

Please note that Appendix 2 gives the designations of pneumatic system components.

Chapter 2 | Elimination of Signal Conflicts by Idle-return Rollers

An idle-return roller pneumatic valve, as shown in Figure 2.1, consists of a 3/2-DC valve fitted with an idle-return roller mechanism. The valve is mechanically actuated by the cam of the associated cylinder just before the cylinder reaches its end position in a particular direction. The valve then gets released when the cylinder reaches its end position. That means the valve generates a short signal pulse just before the cylinder reaches its end position. During the cylinder's movement in the other direction, the idle-return mechanism rolls over without actuating the valve. That means the valve generates no signal in the idle-return direction.

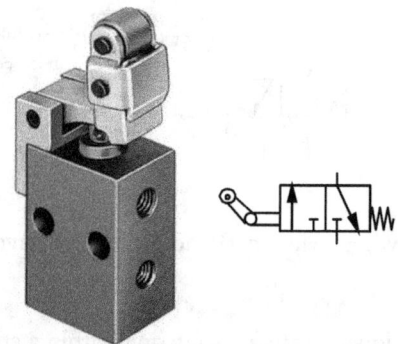

Figure 2.1 | Symbol of idle-return roller

Control Task 2.1 | A Pneumatically-controlled Stamping System
The stamping operation uses a job-positioning cylinder A (1.0) and a stamping cylinder B (2.0). Cylinder A (1.0) extends, bringing the job (workpiece) under cylinder B (2.0). Cylinder B, then, extends and stamps the job. Cylinder A can return only after Cylinder B has retracted fully. A pneumatic control circuit must be developed to implement the control task.

Solution

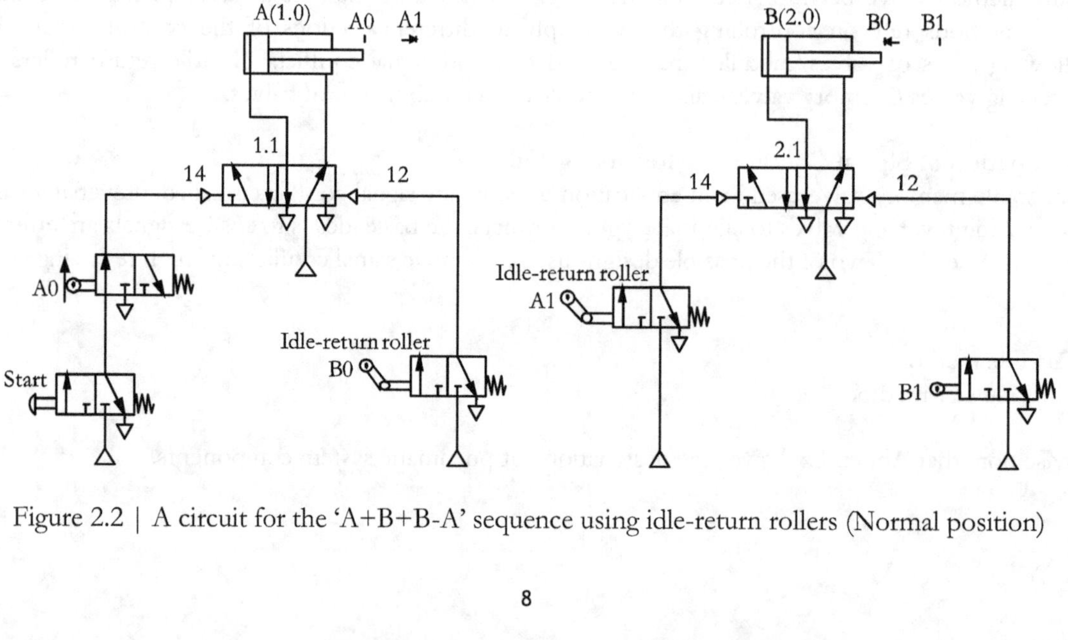

Figure 2.2 | A circuit for the 'A+B+B-A' sequence using idle-return rollers (Normal position)

Figure 2.2 shows the circuit for obtaining the control sequence 'A+B+B-A' using the idle-return rollers fitted at positions B0 and A1. The roller valves at positions A0 and B1 need not be replaced with idle-return rollers, as these valves do not cause signal conflicts in the given sequence of cylinder actions. The critical positions of the circuit are given in Figures 2.3(a) through (e).

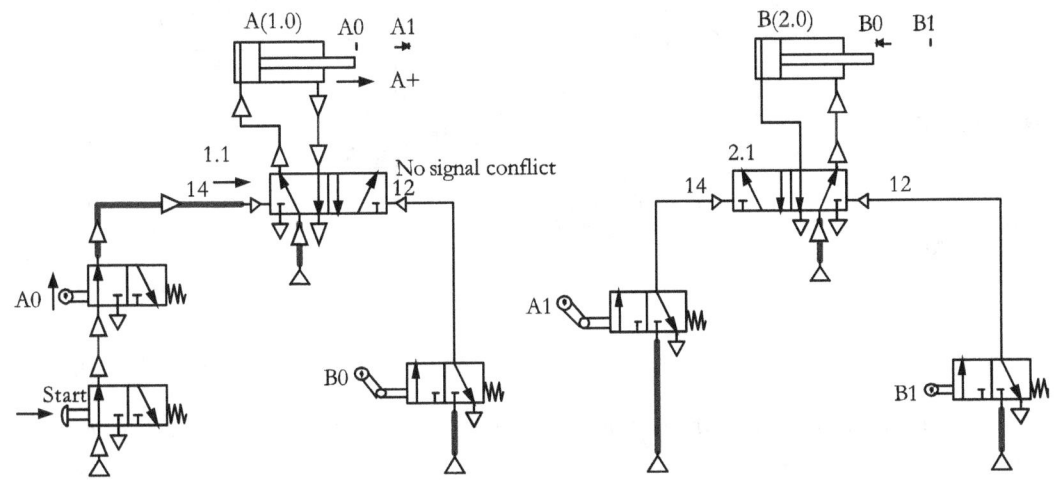

Figure 2.3(a) | The circuit position when the Start pushbutton is pressed (A+ action)

Figure 2.3(a) shows the position when the 'start' pushbutton is pressed. Idle-return roller valve B0 is released in the fully retracted position of cylinder B. There is no signal conflict at valve 1.1. Valve 1.1 then switches over, and cylinder A extends. (A+).

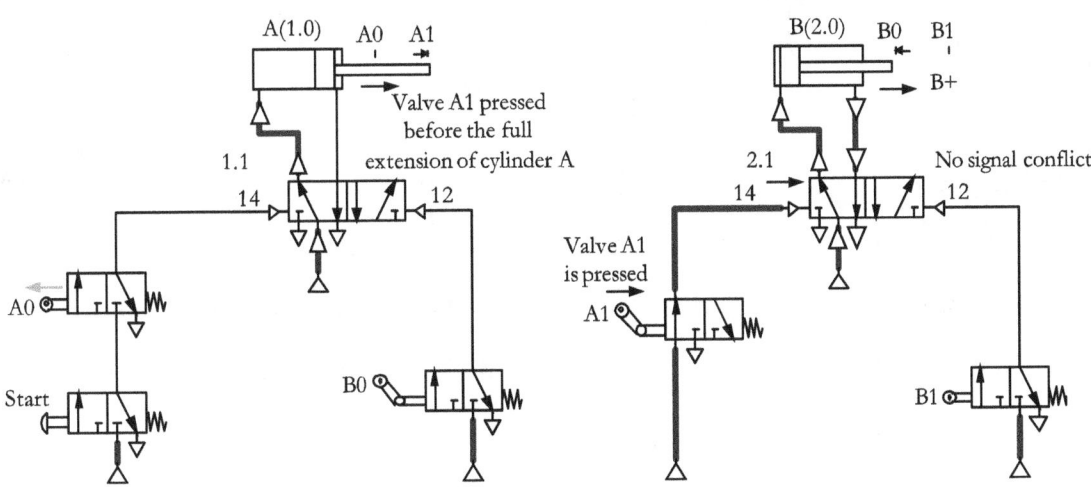

Figure 2.3(b) | The circuit position when idle-return roller valve A1 is pressed (B+ action)

Figure 2.3(b) shows the position when the idle-return roller valve A1 is pressed just before cylinder A reaches its extreme forward end position. There is no signal conflict at valve 2.1, as there is no signal output from valve B1. Valve 2.1 then switches over, and cylinder B extends. (B+)

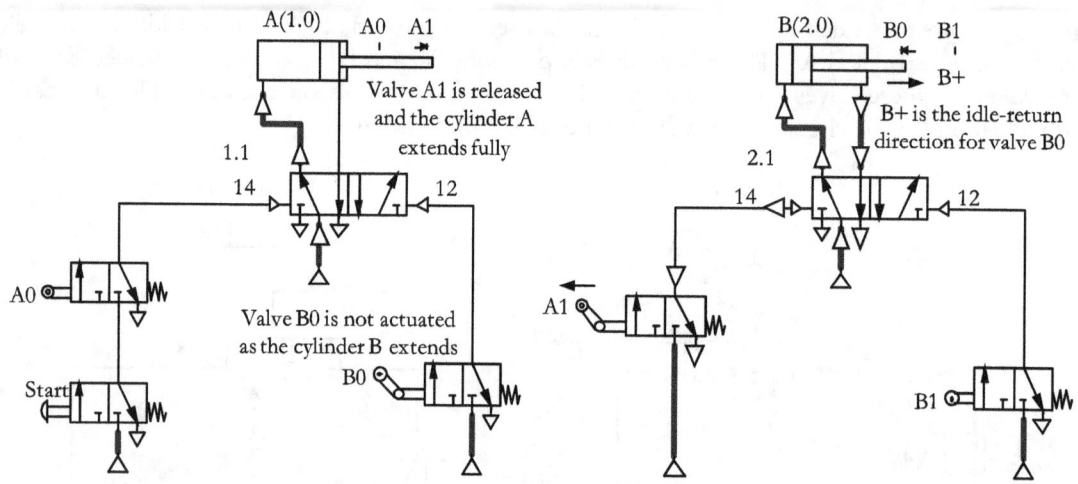

Figure 2.3(c) | The circuit position when the idle-return roller valve A1 is released

Figure 2.3(c) shows the position of the circuit when valve A1 gets released as cylinder A reaches its end position.

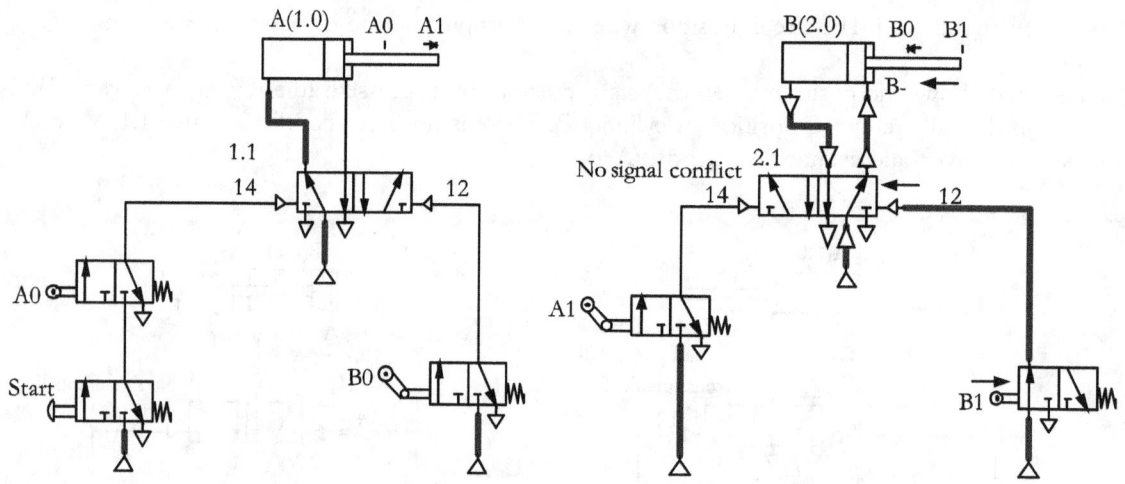

Figure 2.3(d) | The circuit position when roller valve B1 is pressed (B- action)

Figure 2.3(d) shows the position when valve B1 is pressed. The idle-return roller valve A1 is released in the extreme forward end position of cylinder A. There is no signal conflict at valve 2.1. Valve 2.1 then switches over, and cylinder B retracts. (B-)

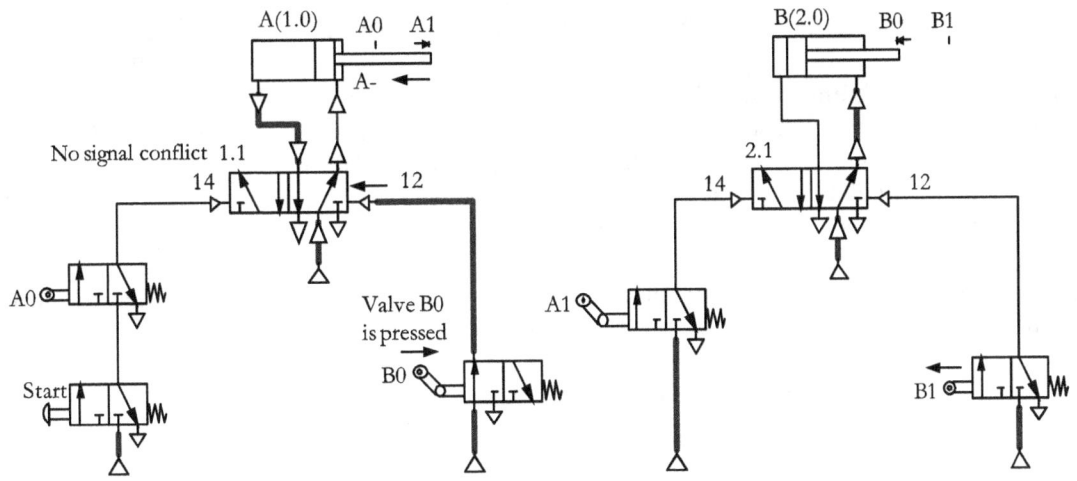

Figure 2.3(e) | The circuit position when idle-return roller valve B0 is pressed (A- action)

Figure 2.3(e) shows the position when the idle-return roller valve B0 is pressed just before cylinder B reaches its fully retracted position. There is no signal conflict at valve 1.1, as there is no signal at the other end of the valve. Valve 1.1 then switches over, and cylinder A retracts. (A-)

Valve B0 gets released as cylinder B reaches its fully retracted position.

Conclusion
Figures 2.3(a) through 2.3(e) show that using two idle-return roller valves (A1 and B0) eliminates signal conflicts, ensuring proper sequential operation.

Drawbacks of Idle-return Rollers
The use of idle-return rollers for eliminating signal conflicts in pneumatic circuits has the following disadvantages:
- This method is not reliable.
- The end position cannot be sensed accurately.
- Fast control systems cannot be set up.

Hence, idle-return rollers are not used in practice in multi-cylinder circuits. The use of idle-return rollers may be justified only for simpler circuits.

Chapter 3 | Cascade Method – Design and Operating Principles

Control Task 3.1 | A Pneumatically-controlled Stamping System
The stamping operation uses a job-positioning cylinder A (1.0) and a stamping cylinder B (2.0). Cylinder A (1.0) extends, bringing the job (workpiece) under cylinder B (2.0). Cylinder B, then, extends and stamps the job. Cylinder A can return only after cylinder B has retracted fully. A pneumatic control circuit must be developed to implement the control task.

Figures 3.1 and 3.2 show a circuit for the sequence of operation A+B+B-A- of cylinders A and B in two critical positions. Figure 3.1 shows the circuit's position when a signal conflict occurs at valve 1.1.

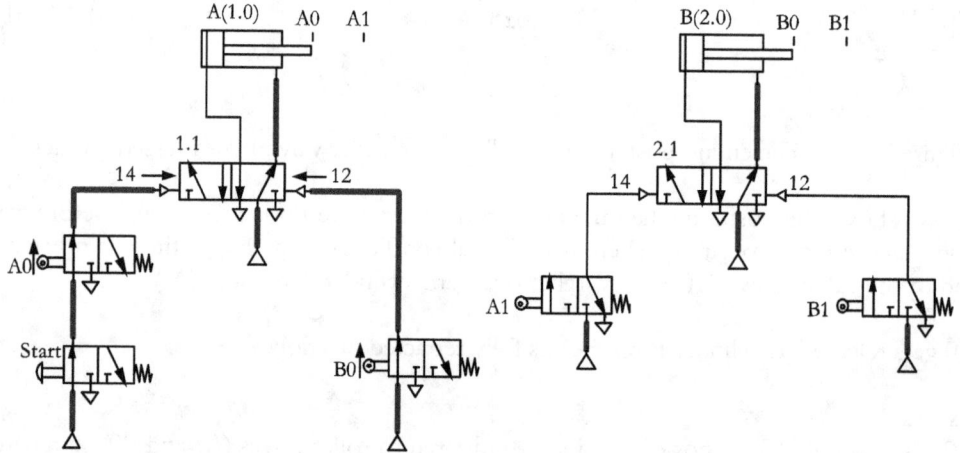

Figure 3.1 | Signal conflict at valve 1.1

Figure 3.2 shows the circuit's position when a signal conflict occurs at valve 2.1. A careful examination of the circuit reveals that signal conflicts occur when compressed-air signals reach both ends of the valve simultaneously.

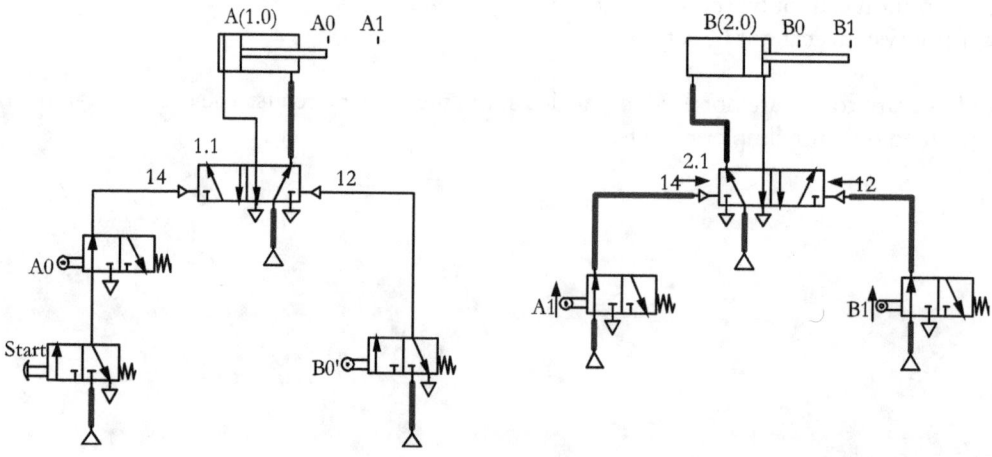

Figure 3.2 | Signal conflict at valve 2.1

Cascade Method

A possible solution to eliminate signal conflicts is to prevent signals from appearing simultaneously at both ends of each valve. This idea is used in the cascade method. The most critical steps for developing a pneumatic circuit for the control task are: (1) Dividing the given work operations into a certain number of groups, and (2) Dividing the power supply into the same number of groups.

Grouping of Work Operations

In this method, first, the sequence of operations (A+B+B-A-) for cylinders A and B in control task 3.1 can be divided into an appropriate number of groups to avoid signal conflicts.

If A+ and A- operations happen to fall in one group, then there is a possibility of signals appearing simultaneously at both ends of the final control element controlling cylinder A. As a consequence, there is a possibility of a signal conflict.

The same is true for B+ and B- operations. Hence, the sequence of operations should be divided so that operations A+ and A- fall into different groups, and B+ and B- fall into different groups. Such a grouping is shown in Figure 3.3.

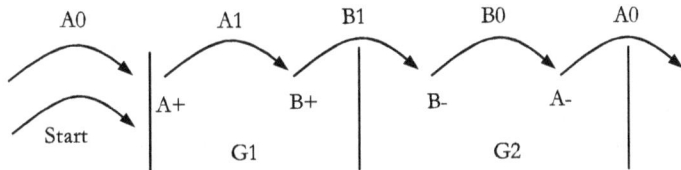

Figure 3.3 | Grouping of operations

As shown in the Figure, the first two actions, A+ and B+, are placed in group G1. The next two actions, B- and A-, if placed in group G1, may cause signal conflicts; hence, they are placed in group G2. Remember, the desired sequence of actions should be maintained.

Another factor to keep in mind is that as the number of groupings increases, the number of valves required for the grouping circuit also increases. Hence, in this method, every attempt should be made to keep the number of groupings to a minimum.

Dividing the Power Supply into Groups

After dividing the sequence of cylinder operations into many groups, the next requirement of the cascade method is to divide the power supply for the control circuit into the same number of groups as for the sequence of operations, in such a way that at any given point of time, only one group will have a supply and all other groups are connected to the exhaust.

By appropriately connecting double-pilot 5/2-DC valves (or reversing valves or memory valves) in series, the power supply can be divided into two, three, four, or more groups. These groupings are illustrated with multiple power-supply positions to facilitate analysis in the following sections. Here, e1, e2, e3, etc. are the input signals to the cascade, and G1, G2, G3, etc. are the output signals from the cascade. The basis of the cascade principle is the precise relationship between inputs e1, e2, e3, etc., and outputs G1, G2, G3, etc. Carefully study the standard group-changing circuits for two, three, and four groups illustrated in the following sections for a deeper understanding.

Group Changing Circuit for Two Groups

The critical positions of a two-group circuit are illustrated in Figure 3.4. A pilot-operated 5/2 DC valve (reversing valve) is used to realize the group-changing circuit. Initially, the power supply should remain in the last group G2, as shown in Figure 3.4(b). When a control signal is applied to the input e1 of the group changing valve, the power supply changes to group G1 from group G2, as shown in Figure 3.4(a). When a control signal is applied to the input e2 of the valve, the power supply changes to group G2 from group G1, as shown in Figure 3.4(b).

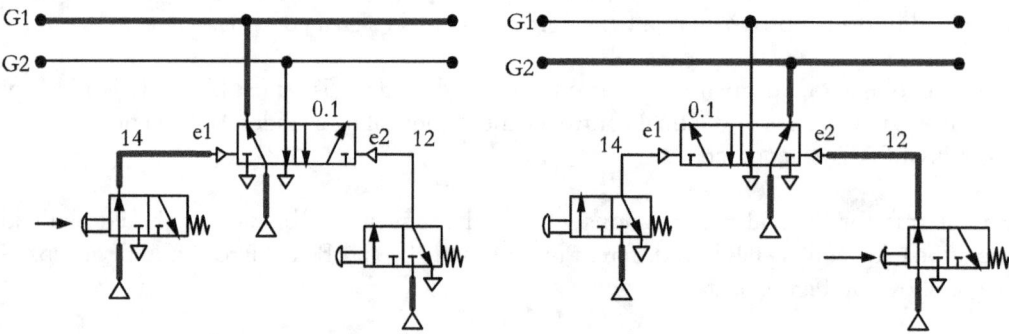

Figure 3.4 | Different positions of a two-group circuit

Group Changing Circuit for Three Groups

Different power-supply positions for a three-group circuit are illustrated in Figure 3.5. Two pilot-operated 5/2 DC valves (reversing valves) are used to realize the group-changing circuit. Initially, the supply is in the last group, G3, as shown in Figure 3.5(c).

When a control signal is applied to the input e1 of the group-changing circuit, the power supply changes to group G1 from group G3, as shown in Figure 3.5(a). When a control signal is applied to the input e2 of the group-changing circuit, the power supply changes to group G2 from group G1, as shown in Figure 3.5(b). When a control signal is applied to the input e3 of the group-changing circuit, the power supply changes to group G3 from group G2, as shown in Figure 3.5(c).

Therefore, when the control signals are applied to inputs e1, e2, and e3 in that sequence, the supply changes to groups G1, G2, and G3, respectively, across the cascade.

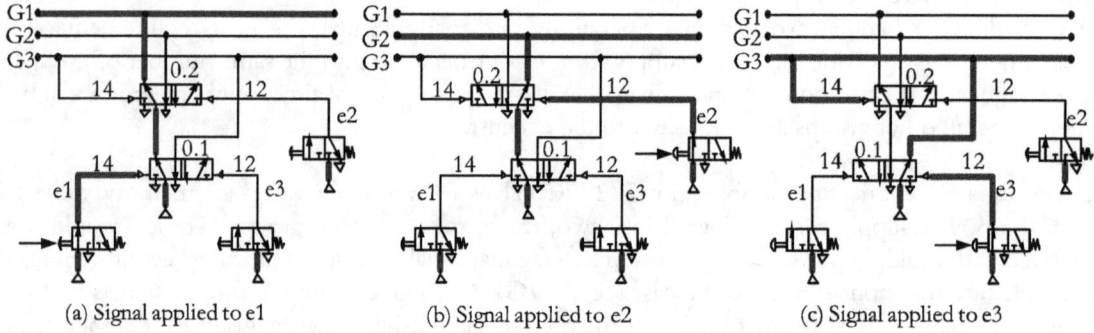

Figure 3.5 | Different positions of a three-group circuit

Group Changing Circuit for Four Groups

Different positions of a group-changing circuit for four groups are illustrated in Figure 3.6. Three pilot-operated 5/2 DC valves (reversing valves) are used to realize the group-changing circuit. Initially, the supply is in the last group, G4, as shown in Figure 3.6(d).

When a control signal is applied to the input e1 of the group-changing circuit, the power supply changes to group G1 from group G4, as shown in Figure 3.6(a). When a control signal is applied to input e2 of the group-changing circuit, the power supply changes to group G2 from group G1, as shown in Figure 3.6(b). When a control signal is applied to the input e3 of the group-changing circuit, the power supply changes to group G3 from group G2, as shown in Figure 3.6(c). When a control signal is applied to the input e4 of the group-changing circuit, the power supply changes to group G4 from group G3, as shown in Figure 3.6(d).

Therefore, when the control signals are applied to the inputs e1, e2, e3, and e4 in that sequence, the supply changes to the groups G1, G2, G3, and G4, respectively, across the cascade.

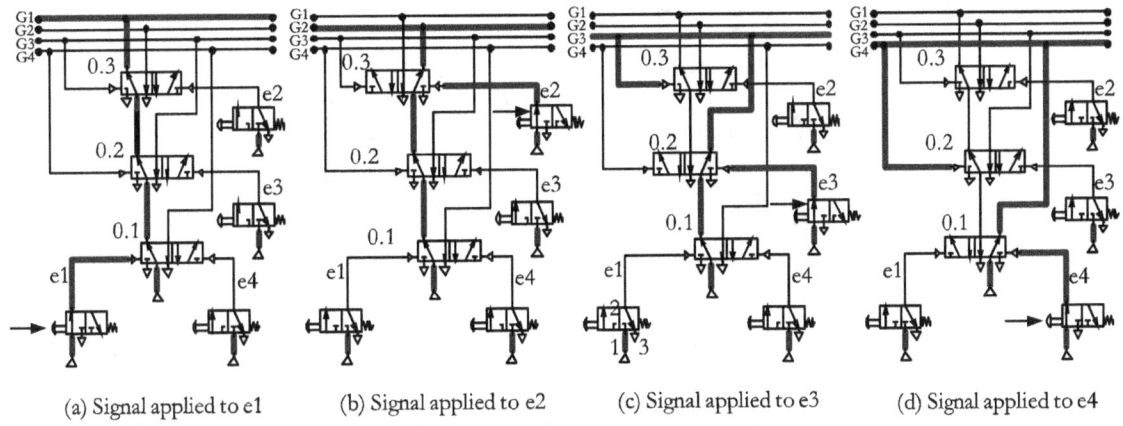

(a) Signal applied to e1   (b) Signal applied to e2   (c) Signal applied to e3   (d) Signal applied to e4

Figure 3.6 | Different power supply positions of a four-group circuit

The cascade circuits can be developed for any number of groups. The arrangement will always remain the same. That is, all valves are connected in series, the first valve at the top issues two output signals, G1 and G2, and all other valves issue one output signal. Each subsequent valve in the series resets the previous one, and so on. The last (bottom) valve in the cascade series receives two input signals.

Circuit Design for a Stamping System Using the Cascade Method (Two Groups)

Control Task 3.1: The stamping operation uses a job positioning cylinder A(1.0) and a stamping cylinder B(2.0). Cylinder A (1.0) extends, bringing the job (work-piece) under cylinder B (2.0). Cylinder B, then, extends and stamps the job. Cylinder A can return only after cylinder B has retracted fully. A pneumatic control circuit must be developed to implement the control task.

Solution
Draw the power circuit showing cylinders A(1.0) and B(2.0), and their corresponding final control elements 1.1 and 2.1, as shown in Figure 3.7. Draw the control circuit with a group-changing circuit for two groups.

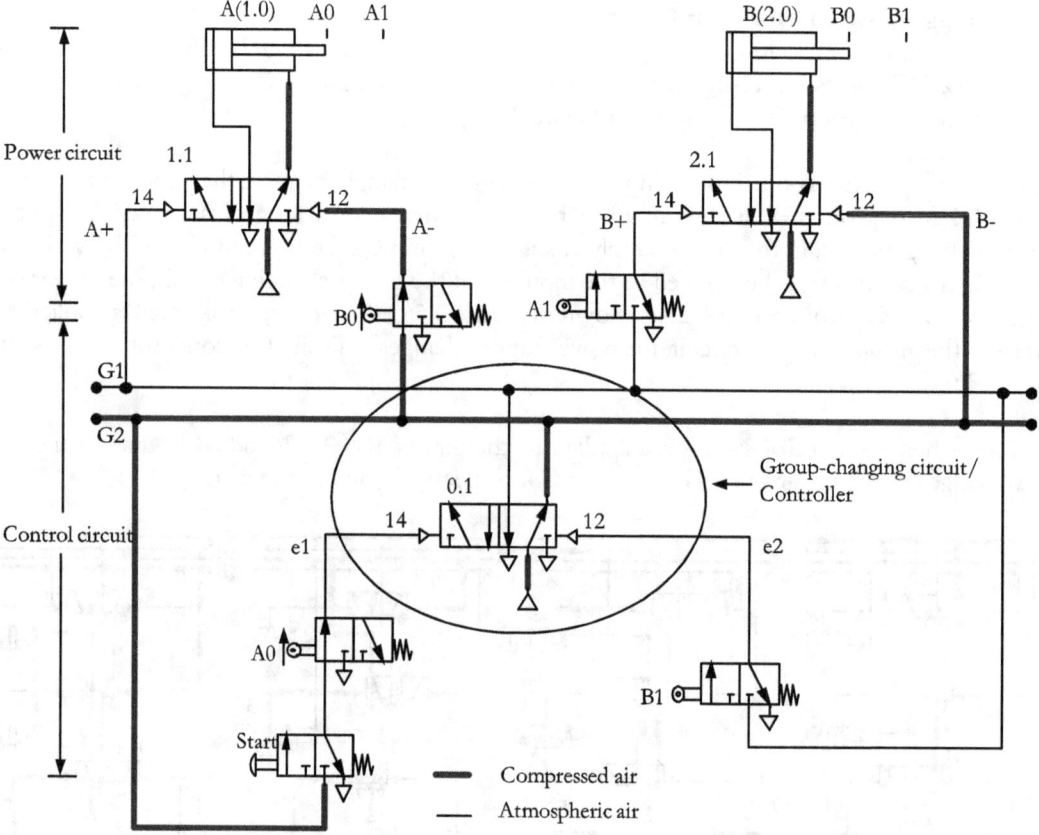

Figure 3.7 | Circuit design using the cascade method (two groups) for control task 3.1

Add the group changing circuit for two groups just below the power circuit. The group-changing circuit serves as the controller, ensuring that only one group receives compressed air at any given time, while all other groups are connected to the exhaust.

Add the control valves/roller valves (sensors) as specified in the notational form given in Figure 3.8. The sensors are used to get an automatic or semi-automatic operation of the system. Initially, sensors A0 and B0 are shown in the actuated position. It may be observed that port 14 (for A+ action) and port 12 (for A- action) of valve 1.1 are always connected to different supply groups. Similarly, ports 14 (for B+ action) and 12 (for B- action) of valve 2.1 are always connected to different supply groups. This prevents signals from appearing simultaneously on both sides of each final control element, thereby avoiding signal conflicts.

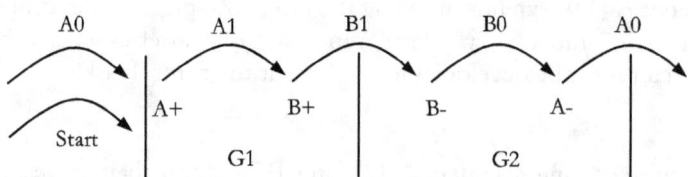

Figure 3.8 | Notational form for control task 3.3

Explanation of the Cascade Circuit for Control Task 3.1

When the 'Start' pushbutton is pressed, the compressed air supply from group G2 is directed to port 14 of reversing valve 0.1 through the actuated 'Start' pushbutton and sensor A0. The reversing valve switches over, causing the group supply to change from G2 to G1. (A0 is shown in the released position due to the subsequent extension of cylinder A)

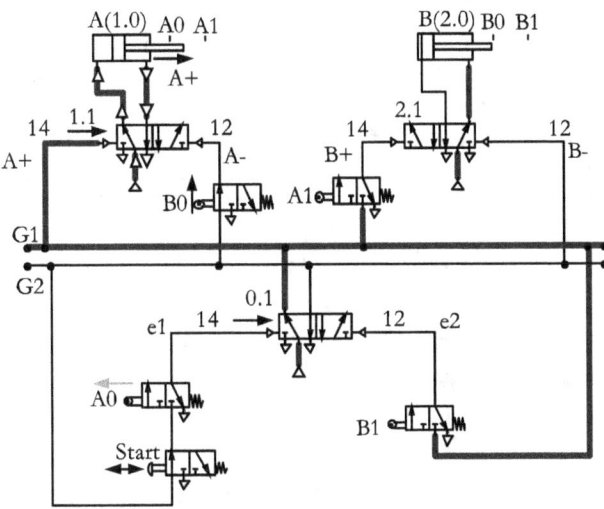

Figure 3.9(a) | A circuit position during A+ action

When the group is changed from G2 to G1 with G2 vented, the air supply from group G1 is directed to port 14 of valve 1.1. As there is no possibility of a signal conflict, valve 1.1 switches over. Hence, cylinder A moves forward. (A+ action) [Figure 3.9(a)].

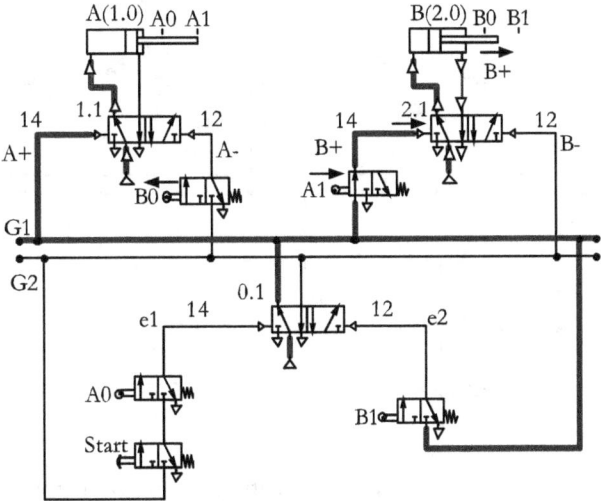

Figure 3.9(b) | A circuit position during B+ action

Sensor A1 is actuated as a result of the A+ action, allowing the air supply from group G1 to reach port 14 of valve 2.1. As there is no signal conflict, valve 2.1 switches over. Hence, cylinder B moves forward automatically. (B+ action) [Figure 3.9(b)].

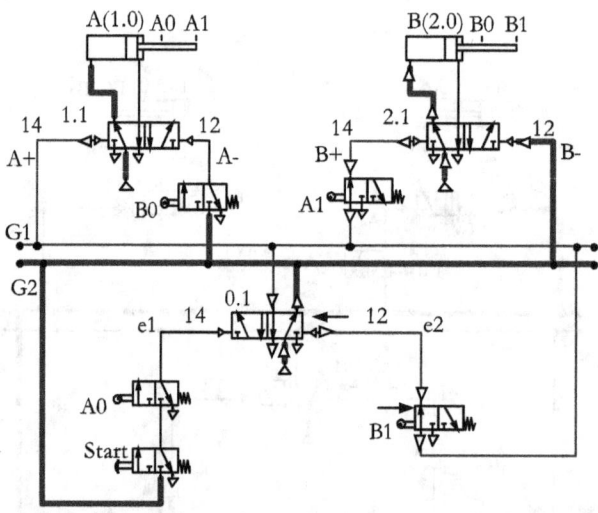

Figure 3.9(c) | Circuit position when changing the group supply from G1 to G2

Sensor B1 is actuated by the B+ action, allowing the air supply from group G1 to reach port 12 of reversing valve 0.1. As a result, the reversing valve switches over, causing the group supply to change from G1 to G2. [Figure 3.9(c)]

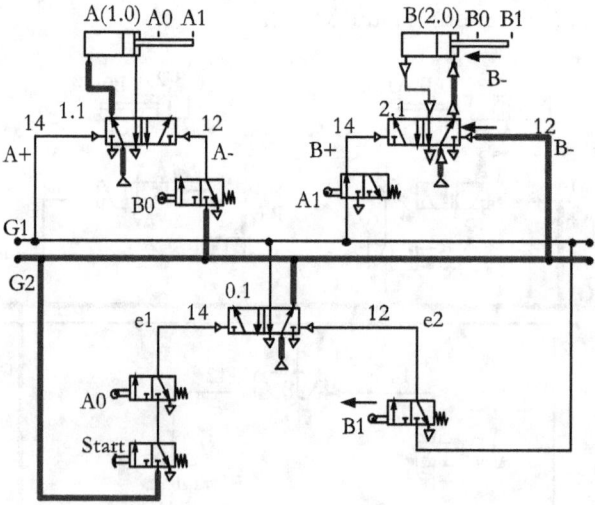

Figure 3.9(d) | A circuit position during B- action

As the group changes from G1 to G2, with G1 vented, the compressed air supply from G2 is directed to port 12 of valve 2.1. As there is no signal conflict, valve 2.1 switches over. Hence, cylinder B retracts automatically. (B- action) [Figure 3.9(d)]

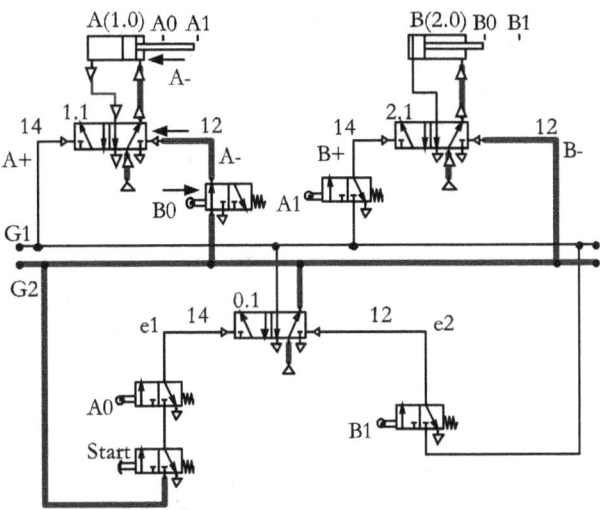

Figure 3.9(e) | Circuit position during A- action

Sensor B0 is actuated as a result of the B- action, allowing the air supply from group G2 to reach port 12 of valve 1.1. As there is no signal conflict, valve 2.1 switches over. Hence, cylinder A retracts automatically. (A- action) [Figure 3.9(e)]

Points for the Development of Pneumatic Circuits Using the Cascade Method
The cascade system provides a straightforward method of designing any sequential circuit. This system will always give a workable circuit, and only rarely will it be possible to suggest any improvements. The cascade system, however, must be carefully monitored by checking the following points:

Preset: The system must be set to the last supply group for the start-up.

Pressure drop: Because the power supply is 'cascaded', a large circuit can suffer from excessive pressure drops.

Cost: Additional valves and connections significantly increase hardware costs.

Limitations of the Cascade Method
A major limitation of the cascade circuit is that the control circuit receives its input power through a single connection. Thus, the compressed air may have to pass through the series-connected valves in the cascade before a control action can be initiated. This flow results in a significant pressure drop when a large number of valves are connected in series, and the cascade circuit's response to input sensor signals tends to slow. Thus, in practice, the number of groups is limited to four or five output signals (three or four reversing valves).

Circuit Design for a Pneumatically-controlled drilling machine using the Cascade Method (three groups)

Control Task 3.2 | Pneumatically-controlled drilling machine

Rectangular prefabricated workpieces are drilled using a pneumatically controlled drilling machine shown in Figure 3.10. The work-pieces are removed from a hopper feed unit. They are pushed and clamped using a clamping cylinder (A), moved slowly by a drill-attached cylinder (B) during drilling, and ejected by an ejecting cylinder (C). The displacement-step diagram is shown in Figure 3.11. The sequence of operations must be carried out either for a single cycle or for continuous cycles using 'start' and 'stop' controls. Develop a pneumatic control circuit to implement the given control task.

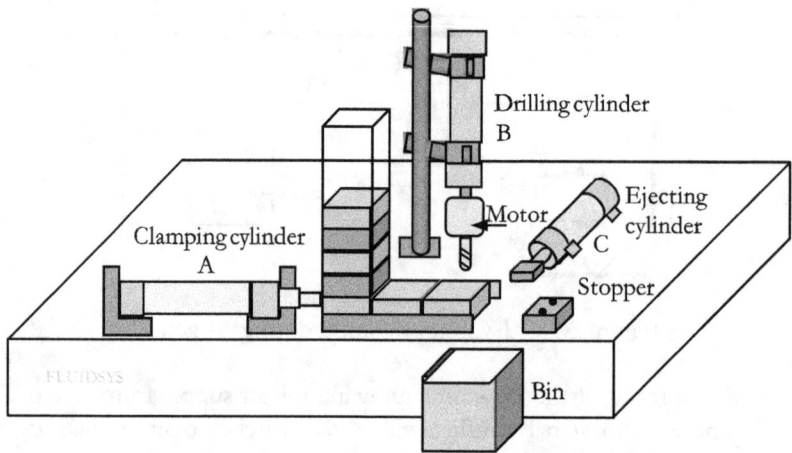

Figure 3.10 | The positional layout of a pneumatically-controlled drilling machine

Displacement-Step Diagram

The displacement-step diagram for control task 3.2 is shown below.

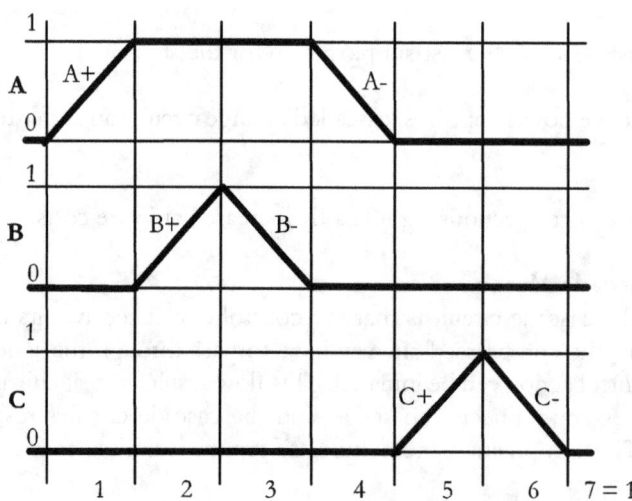

Figure 3.11 | Displacement-step diagram

Solution

The steps to develop a control circuit for the pneumatically-controlled drilling machine specified in control task 3.2, using the cascade method, are given below.

Represent the required control task in the notational form and then divide the sequence of operations into different groups as given in Figure 3.12. Also, show the pushbutton and sensor signals in the Figure.

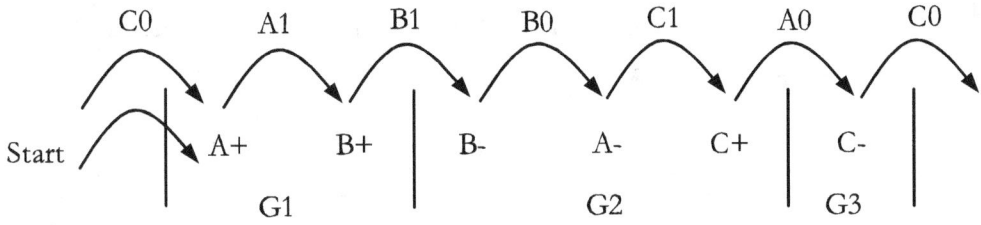

Figure 3.12 | Notational representation of control task 3.2

This grouping for the sequence of operations results in three supply groups: G1, G2, and G3. Operations A+ and B+ take place in group G1, operations B-, A-, and C+ take place in group G2, and operation C- takes place in group G3.

The sensor signals A0, A1, B0, B1, C0, and C1 are used to operate the control sequence automatically. C0 and 'Start' signals are used to obtain the first cylinder operation, A+, via the group change from G3 to G1.

Note: The last cylinder operation, C-, can be assigned to group G1, resulting in only two groups. The resulting circuit will generate a supply in the first group, G1, at the end of each operational cycle. However, by convention, every circuit should generate a supply in the last group at the end of each cycle of operations for uniformity. Further, it may result in an incorrect sequence of operations if some sensors (here, A1 or B1) are inadvertently operated in the starting position.

Draw cylinders A, B, and C and associated final control elements. Mark the sensor positions with appropriate designations. Complete the power circuit and designate the components, preferably with alphabets, for the draft.

Draw the cascade circuit for three groups using two reversing valves just below the power circuit.

Integrate the power circuit and group-changing circuit, and incorporate the pushbuttons and sensors as per the control scheme given in the notational form. The first action in each group is signaled directly by that group's supply.

Incorporate all necessary auxiliary conditions, if any.

Designate all system components using the number system. (Appendix 2 gives the designations of pneumatic system components.)

The cascade circuit for the problem given in control task 3.2 is shown in Figure 3.13.

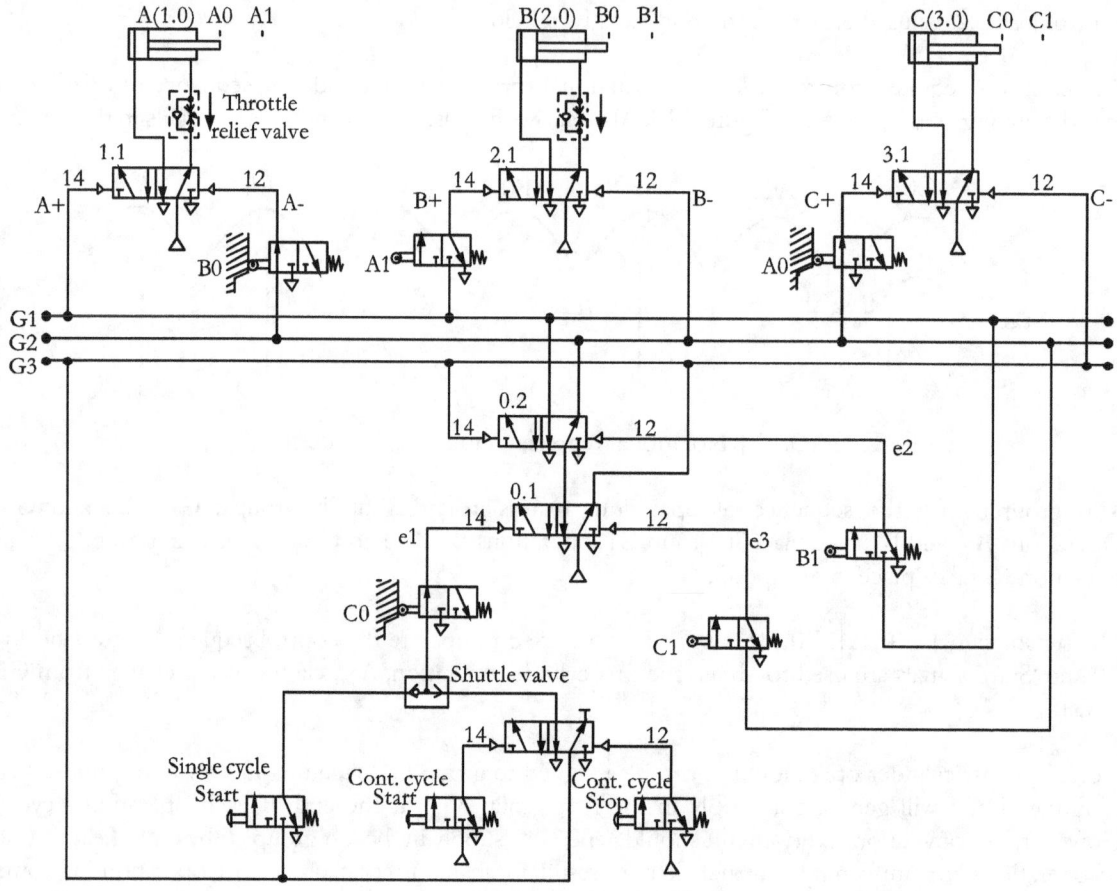

Figure 3.13 | A cascade circuit for the sequence of operations: A+B+B-A-C+C-

Critical positions of the circuit for control task 3.2 are given in Figures 3.14(a), (b), (c), and (d).

The initial position of the circuit for control task 3.2 is shown in Figure 3.14(a).

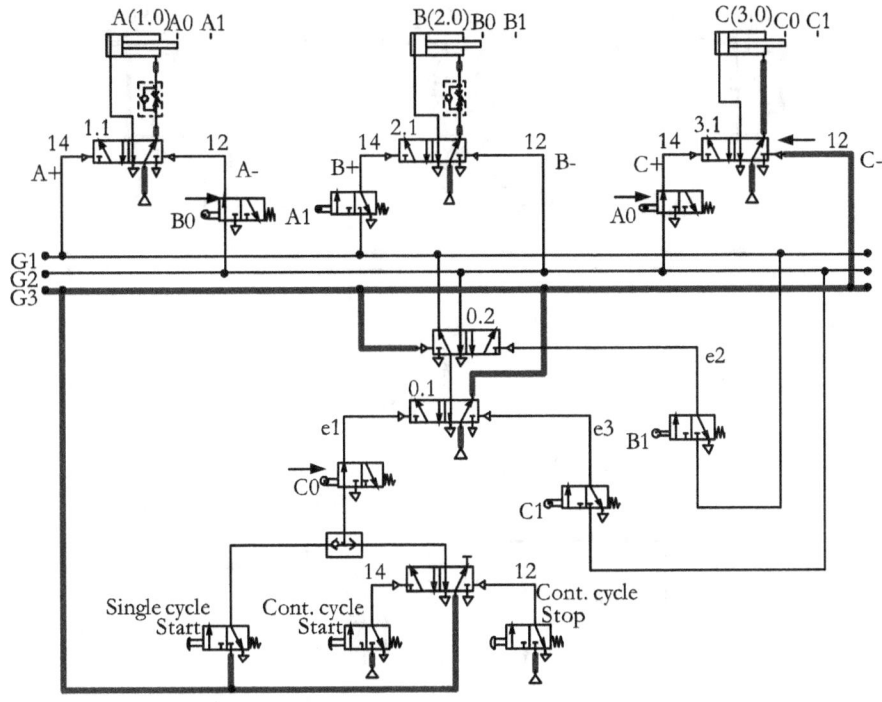

Figure 3.14(a) | Initial position of the circuit

The initial position of the circuit for the sequence of operations (A+ B+ B- A- C+ C-) is shown in Figure 3.14 (a). The control scheme supports both single-cycle and continuous operation. Because there are three groups, two reversing valves (0.1 and 0.2) are used.

All cylinders are supplied with a compressed-air power source. In the initial position, all cylinders are retracted. Each cylinder is controlled by a 5/2-way double-pilot valve (final control element).

As shown, the control supply reaches the last group, G3, through the reversing valve 0.1. Roller valves A0, A1, B0, B1, C0, and C1 act as position sensors.

The critical position of the circuit after A+ and B+ operations is shown in Figure 3.14 (b).

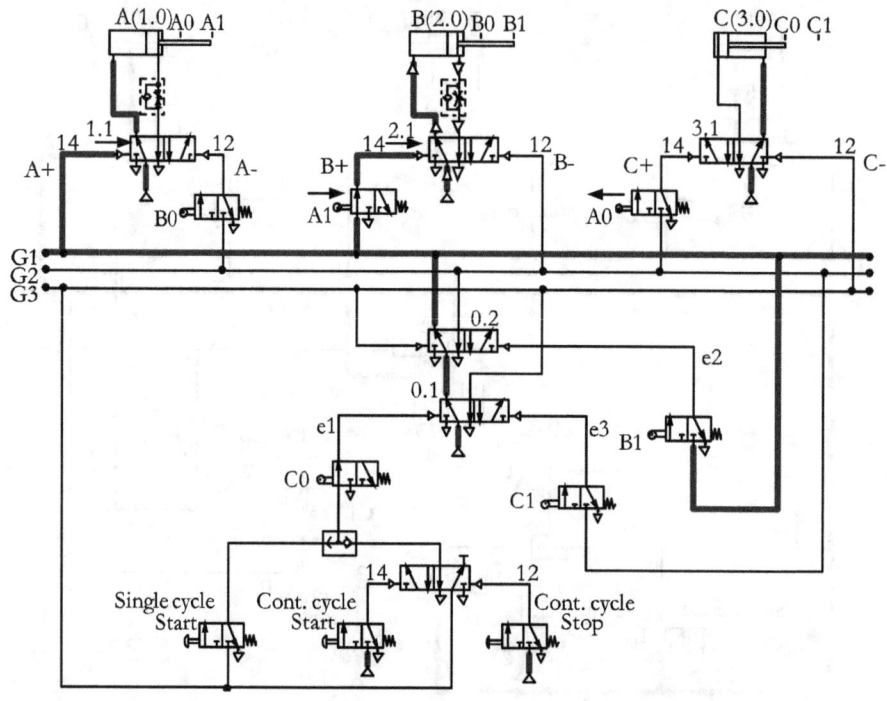

Figure 3.14(b) | The circuit position after A+B+

When a start signal is applied (momentarily or continuously), the reversing valve 0.1 shifts, supplying compressed air to group G1 while exhausting it from group G3.

As soon as the supply reaches group G1, the final control valve 1.1 shifts because there is no signal conflict at the valve. Cylinder A then extends (A+).

When cylinder A is fully extended, roller valve A1 is actuated. As soon as roller valve A1 is actuated, the final control element 2.1 shifts because there is no signal conflict at the valve. Cylinder B then extends (B+).

The critical position of the circuit after A+, B+, B-, A-, and C+ operations is shown in Figure 3.14 (c).

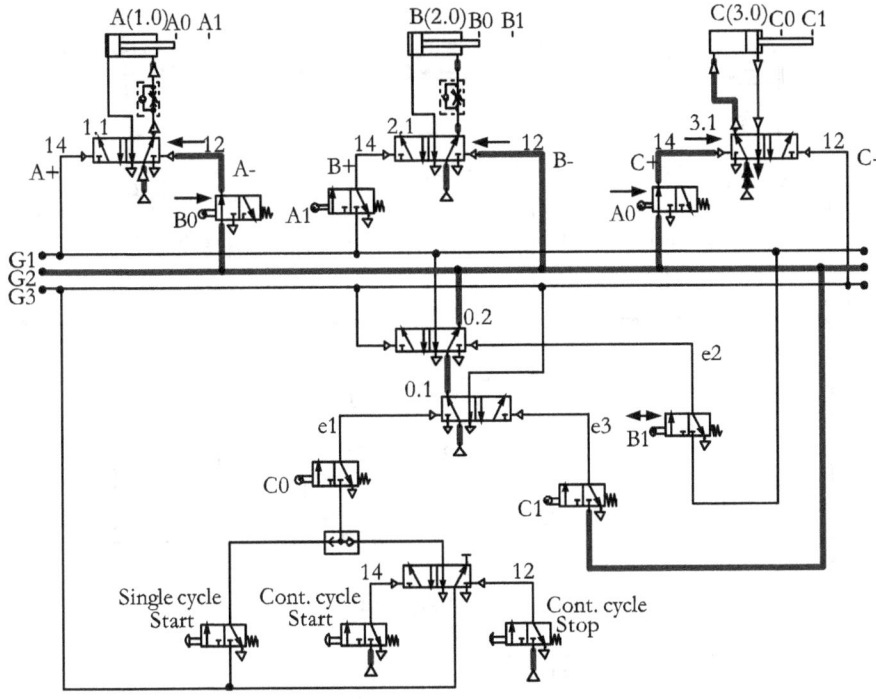

Figure 3.14(c) | The circuit position after A+B+B-A-C+

When cylinder B is fully extended, roller valve B1 is actuated. As soon as roller valve B1 is actuated, the reversing valve 0.2 shifts, supplying compressed air to group G2, while exhausting it from group G1.

As soon as the supply reaches group G2, the final control valve 2.1 shifts because there is no signal conflict at valve 2.1. Cylinder B then retracts (B-).

When cylinder B is fully retracted, roller valve B0 actuates, shifting final control element 1.1. Cylinder A then retracts (A-). When cylinder A is fully retracted, roller valve A0 actuates, shifting final control element 3.1. Cylinder C then extends (C+).

The critical position of the circuit after A+, B+, B-, A-, C+, and C- operations is shown in Figure 3.14 (d).

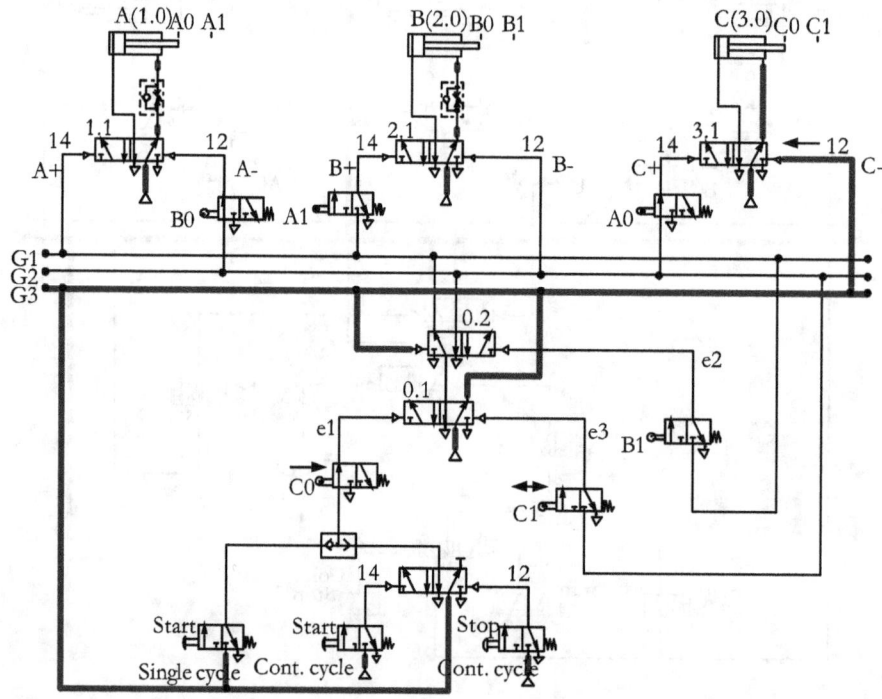

Figure 3.14(d) | The circuit position after A+B+B-A-C+C-

When cylinder C is fully extended, roller valve C1 is actuated. As soon as roller valve C1 is actuated, the reversing valve 0.1 shifts, supplying compressed air to group G3, while exhausting it from group G2.

As soon as the supply reaches group G3, the final control valve 3.1 shifts because there is no signal conflict at valve 3.1. Cylinder C then retracts (C-).

Note: If a continuous cycle start is applied, the sequence of operations will repeat until a continuous cycle stop is applied.

It can also be observed that the cyclic operation is free of signal conflicts.

Control Task 3.3 | Pneumatically-controlled Processing Machine to Transfer Parts

Parts are to be supplied to a processing machine from the discharge rail of a vibrator conveyor, as shown in the schematic diagram in Figure 3.15. The machine is provided with fixtures to receive the parts. Cylinder A(1.0) extends and picks a job part from the discharge rail. Cylinder A then retracts with the part. Cylinder B (2.0), which is hinged to cylinder A, can push the entire cylinder A, along with the part, to align with the fixture. Cylinder A again extends and transfers the part into the fixture. Cylinder A and cylinder B then retract to their respective initial positions. The displacement-step diagram is shown in Figure 3.16. Develop a pneumatic control circuit to implement the control task.

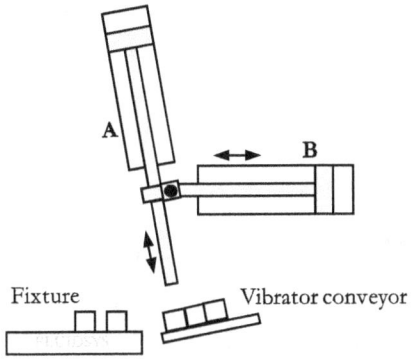

Figure 3.15 | A schematic diagram of a pneumatically-controlled processing machine to transfer parts

Displacement-step diagram

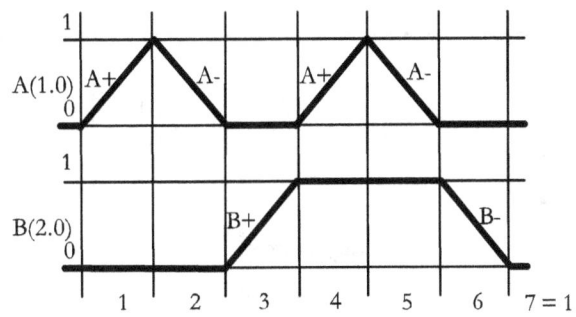

Figure 3.16 | Displacement-step diagram (Control task 3.3)

Solution

The abbreviated notation for control task 3.3, along with the associated signals, is shown in Figure 3.17.

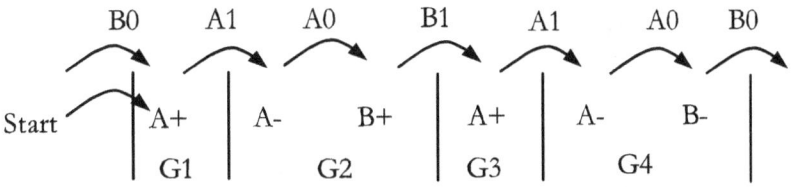

Figure 3.17 | Notational representation of the problem given in control task 3.3

From the notational representation of the control task, it can be observed that sensors A0 and A1 are actuated twice in each cycle. The signals from these sensors should be directed to the appropriate cascade inputs to obtain the required control sequence, as shown in Figure 3.18. For example, when signal A1 first appears in the sequence, it changes the group from G1 to G2 by applying it to the cascade input e2. Next, when it appears a second time in the sequence, signal A1 is applied to the cascade input e4, changing the group from G3 to G4. This requirement can be achieved by logically combining the signals from A1 and G1 using an AND valve to direct the output to cascade input e2, and by logically combining the signals from A1 and G3 to direct the output to cascade input e4.

Similarly, signals from sensor A0 can be directed to the appropriate inputs.

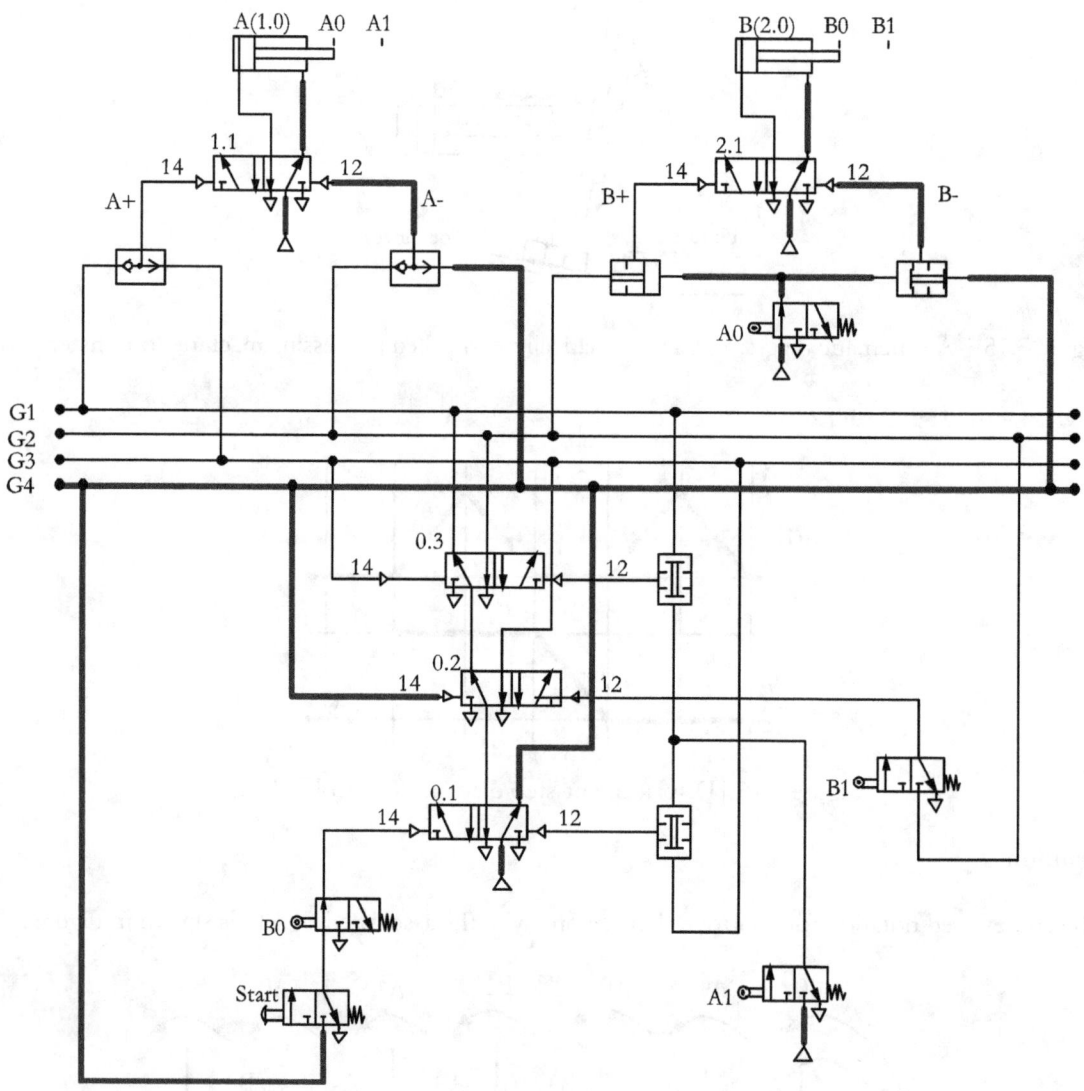

Figure 3.18 | The initial position of the pneumatic circuit for control task 3.3

Control Task 3.4 | Drilling Station

Work-pieces are continuously processed in a drilling station, as shown in Figure 3.19(a), consisting of a clamp cylinder, drill cylinder, and eject cylinder. The work-pieces are stacked in a gravity chute. The clamping cylinder pushes workpieces one after another from the gravity chute and clamps them for drilling. Further, the drilling cylinder feeds the attached drilling machine, which drills the workpieces. The eject cylinder ejects the drilled work-pieces. Before a new workpiece is placed, the drilling area should be briefly sprayed with compressed air. The displacement-step diagram is shown in Figure 3.19(b). Develop a pneumatic control circuit to implement the given control task.

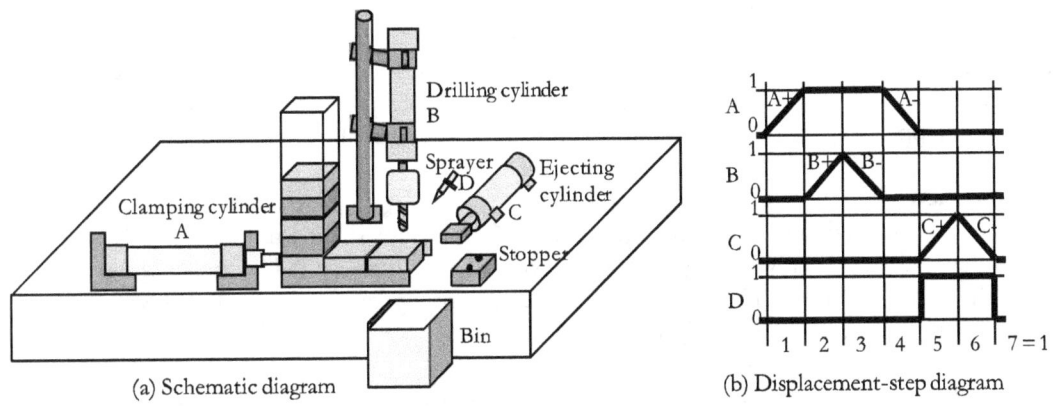

Figure 3.19 | A schematic diagram and displacement step diagram (Control task 3.4)

Solution for A+ B+ B- A- (C+ D+) C- D-
See Figure 3.20

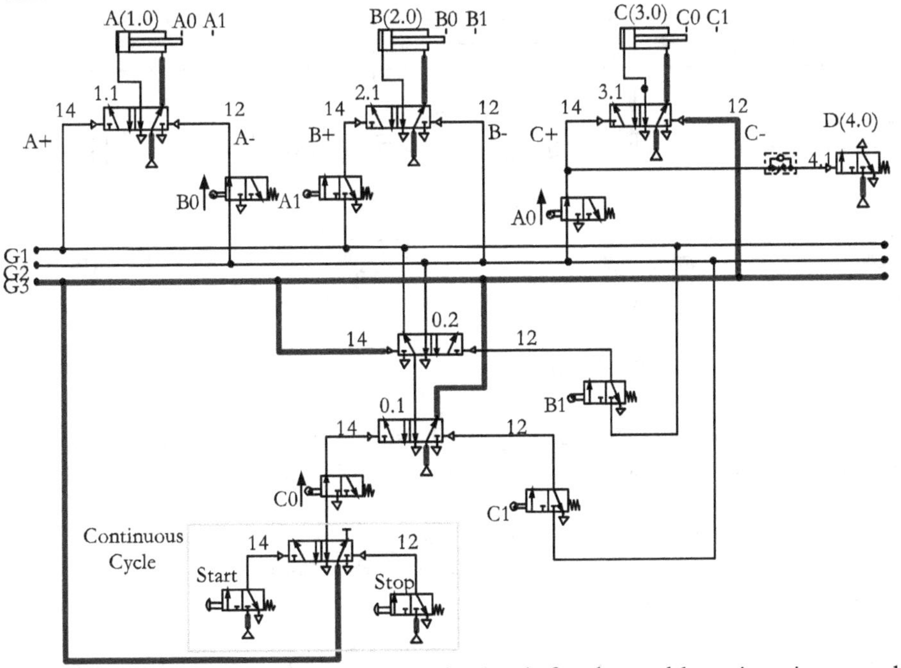

Figure 3.20 | The initial position of the pneumatic circuit for the problem given in control task 3.4

Chapter 4 | Shift Register Method – Design and Operating Principles

Another way to eliminate signal conflicts is to divide the work operations into many groups, based on the number of operations, and use memory valves connected in parallel to control the groups. The number of memory valves used equals the number of groups. This method of eliminating the signal conflict is called the Shift register method. A shift register controller is also called a 'stepper sequencer'.

Shift Register Method

In this method, a memory valve (reversing valve) is set using signals from the system and the signal from the memory valve of the previous group. Once the memory valve is set, the previous stage's memory valve is reset. The output of the memory valve controls an actuator in the pneumatic system. The concept of the stepper sequencer with four groups is explained in the following section. With the same idea, a stepper sequencer for any number of groups can easily be developed.

Stepper Sequencer

The primary circuit used in the stepper sequencer for four groups is shown in Figure 4.1. It consists of four 3/2-way double pilot reversing (memory) valves 0.1, 0.2, 0.3, and 0.4, switches the power supply open for the groups G1, G2, G3, and G4, respectively, using signals X1, X2, X3, and X4, each logically combining with the previous group signal.

Initially, by convention, the memory valve 0.4 for the last group G4 is set, and the supply is maintained in G4. At the same time, all the other groups (G1, G2, and G3) are exhausted. In pneumatic circuits, by convention, the supply should be in the last group. Therefore, the reversing valve for the last group should be set, and all other reversing valves should be reset.

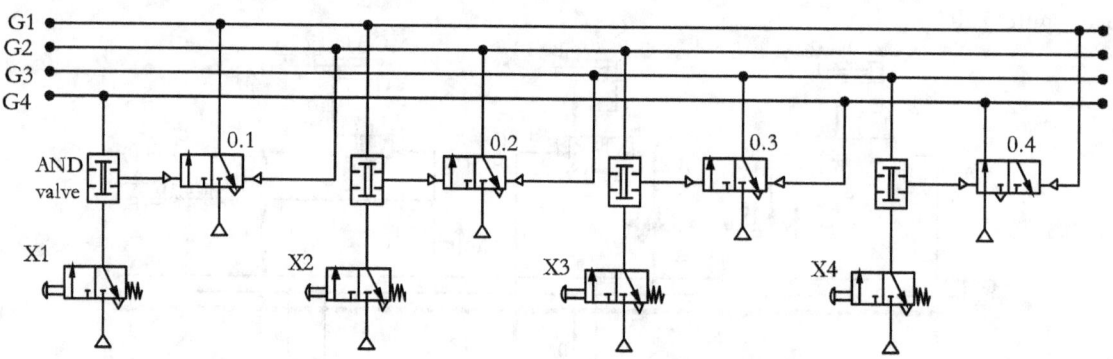

Figure 4.1 | The basic circuit of a shift register (four groups)

Subsequently, the reversing valves 0.1, 0.2, 0.3, and 0.4 sequentially switch on the power supply for the succeeding groups G1, G2, G3, and G4, using signals X1, X2, X3, and X4, respectively.

It can also be observed that when one group is open for power supply, the previous group is exhausted. It can further be observed that there is a definite relationship between inputs x1, x2, x3, and x4, and outputs G1, G2, G3, and G4, respectively.

The four positions of the circuit when signals X1, X2, X3, and X4 are applied one after another in that sequence through pushbuttons are shown in Figures 4.2(a), (b), (c), and (d).

When X1 Signal is Momentarily Applied

Initially, the reversing (memory) valve 0.4 is set, and the supply is in the last group G4, as shown in Figure 4.1. As shown in Figure 4.2(a), signal X1 ANDed with the group G4 sets reversing valve 0.1, generating output G1. As soon as the supply reaches group G1, reversing valve 0.4 resets, exhausting the compressed air from group G4.

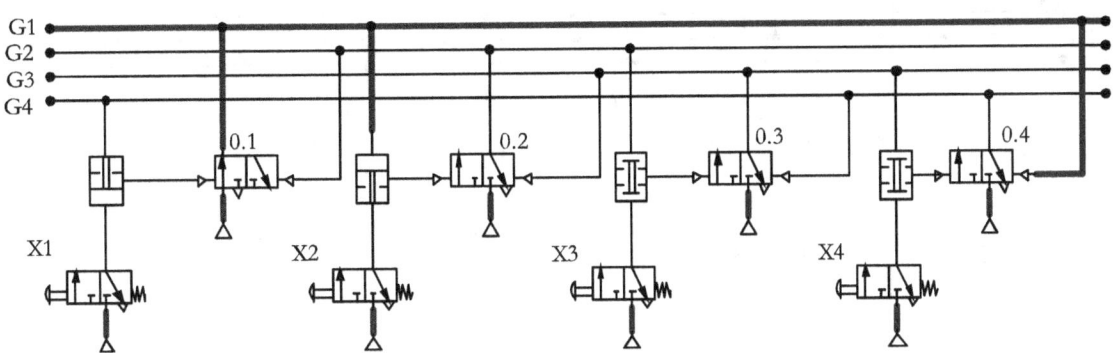

Figure 4.2(a) | x1 -> G1

When X2 Signal is Momentarily Applied

As shown in Figure 4.2(b), signal X2 ANDed with the group G1, sets reversing valve 0.2, generating output G2. As soon as the supply reaches group G2, reversing valve 0.1 resets, exhausting the compressed air from group G1.

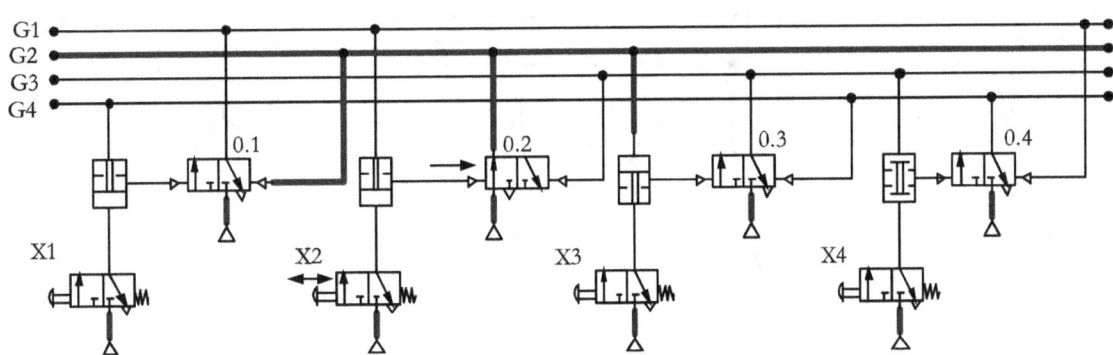

Figure 4.2(b) | x2 -> G2

When X3 Signal is Momentarily Applied
As shown in Figure 4.2(c), signal X3 ANDed with the group G2, sets reversing valve 0.3, generating output G3. As soon as the supply reaches group G3, reversing valve 0.2 resets, exhausting the compressed air from group G2.

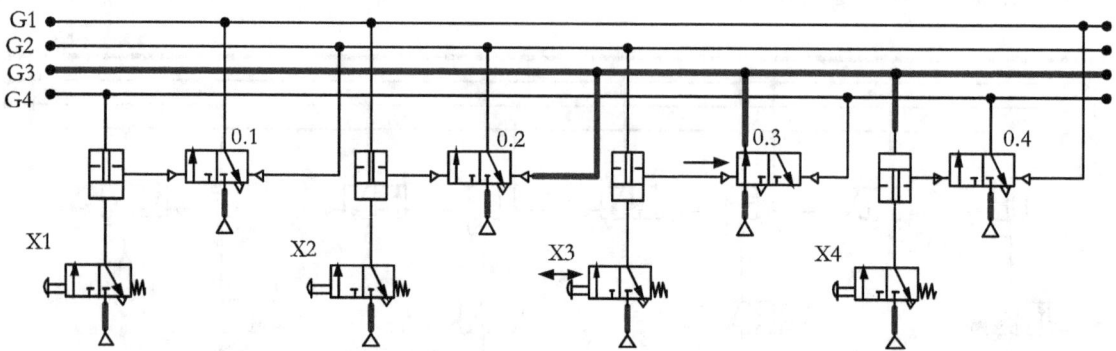

Figure 4.2(c) | x3 -> G3

When X4 Signal is Momentarily Applied
As shown in Figure 4.2(d), signal X4 ANDed with the group G3, sets reversing valve 0.4, generating output G4. As soon as the supply reaches group G4, reversing valve 0.3 resets, exhausting the compressed air from group G3.

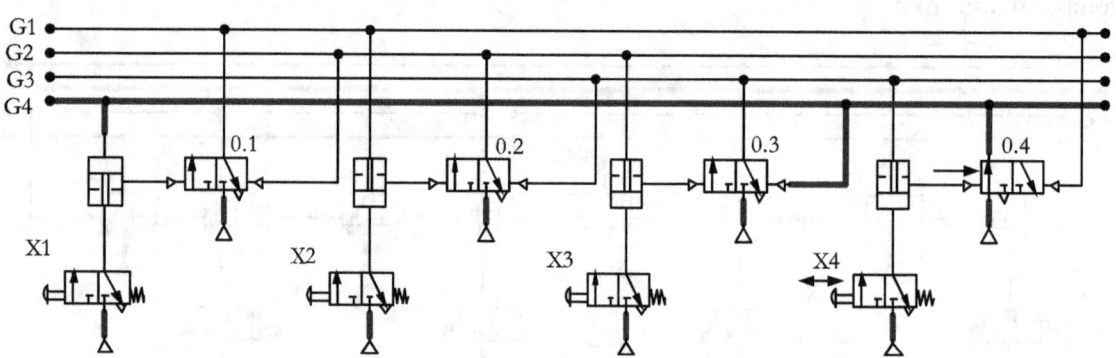

Figure 4.2(d) | x4 -> G4

Integration of Stepper Sequencer into Pneumatic Power Circuit
The concept of the shift register method for eliminating signal conflicts is explained with the problem given in control task 4.1.

The pneumatic stepper sequencer can be integrated into the pneumatic power circuit for cylinders A (1.0) and B (2.0). The critical positions of the circuit for the solution of control task 4.1 are given in Figures 4.4(a) through (d). The initial position of the circuit is given in Figure 4.4(a).

Circuit Design for a Stamping System Using the Shift Register Method (Two Groups)

Control Task 4.1 | A pneumatically-operated Stamping System

The stamping operation uses a job-positioning cylinder A (1.0) and a stamping cylinder B (2.0). Cylinder A (1.0) extends, bringing the job (workpiece) under cylinder B (2.0). Cylinder B, then, extends and stamps the job. Cylinder A can return only after cylinder B has retracted fully. A pneumatic control circuit must be developed to implement this control task using the shift register method.

Solution

The notation for the control task is shown in Figure 4.3. Each group is controlled by a 3/2 (or 5/2, with port 2 blocked) memory valve. Every memory valve is directly connected to the compressed air supply. Thus, the pressure drop across the control elements can be kept low even with a large number of stages. All the memory valves (0.1, 0.2, 0.3, and 0.4) are connected in a parallel arrangement as shown. By convention, initially, the compressed air supply should be in the last group.

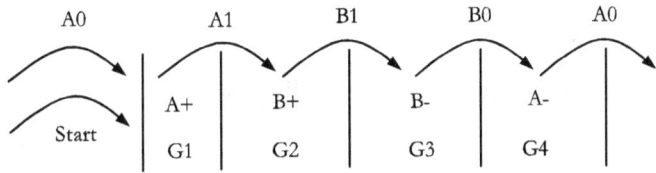

Figure 4.3 | Notational form of representation of the problem in control task 4.1

Control signals (X1, X2, X3, and X4) are applied in that sequence to set the memory valves (0.1, 0.2, 0.3, and 0.4, respectively) through pushbuttons/sensors. Here, the signals are shown applied manually by using pushbuttons. However, it should be noted that actual signals might appear automatically through connected sensors. The control signals are logically combined with the previous group signal via an AND valve, and the AND valve's output is applied to the corresponding memory valve to obtain the correct sequence. Once the memory valve is set, the previous stage's memory valve is reset.

The Initial Position of the Circuit (Control Task 4.1)

Initially, the reversing valve 0.4 is set, and the supply is in the last group G4, as shown in Figure 4.4(a). Cylinders A and B are retracted.

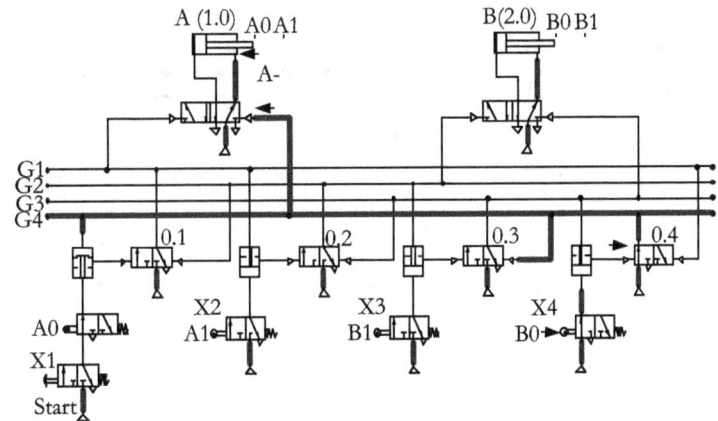

Figure 4.4(a) | The initial position (X4 -> G4) of the circuit for control task 4.1

When Start Signal is Applied

As soon as the 'Start' signal is applied, reversing valve 0.1 is set, and the supply reaches group G1, as shown in Figure 4.4(b). The reversing valve 0.4 is reset, and cylinder A extends (A+). Also, note that there is no signal conflict.

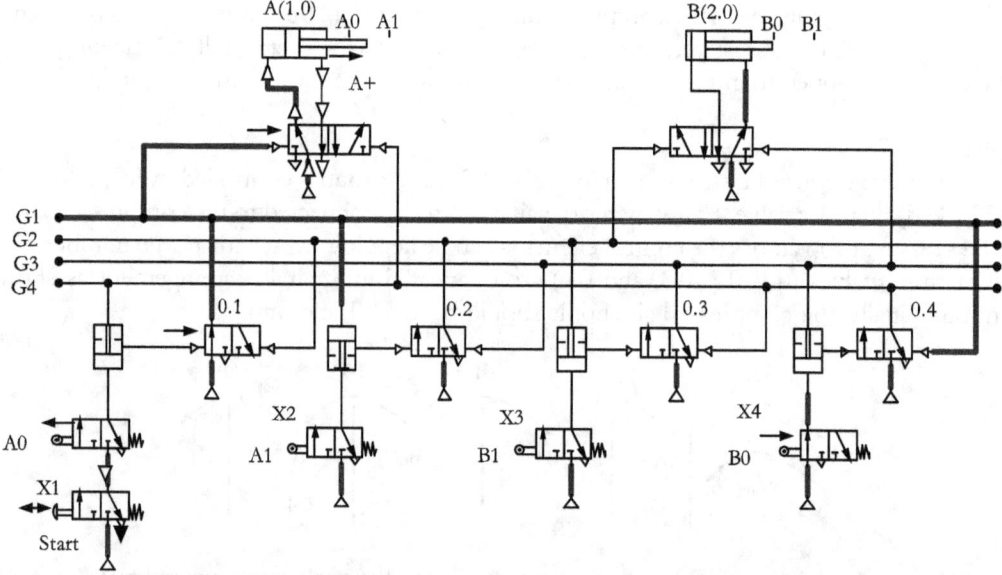

Figure 4.4(b) | The position of the circuit when x1 signal is applied

When Sensor A1 is Actuated

When cylinder A extends, sensor A1 senses. As soon as the sensor A1 signal is applied, reversing valve 0.2 is set, and the supply reaches group G2, as shown in Figure 4.4(c). The reversing valve 0.1 is reset, and cylinder B extends (B+). Also, note that there is no signal conflict.

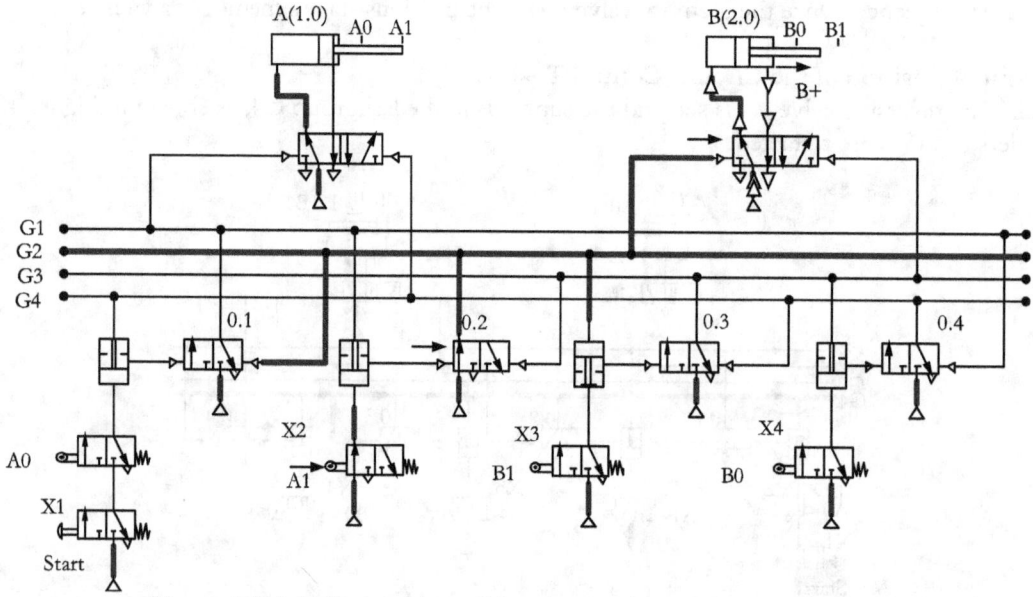

Figure 4.4(c) | The position of the circuit when x2 signal is applied

When Sensor B1 is Actuated

When cylinder B extends, sensor B1 senses. As soon as the sensor B1 signal is applied, reversing valve 0.3 is set, and the supply reaches group G3, as shown in Figure 4.4(d). The valve 0.2 is reset, and cylinder B retracts (B-). Also, note that there is no signal conflict.

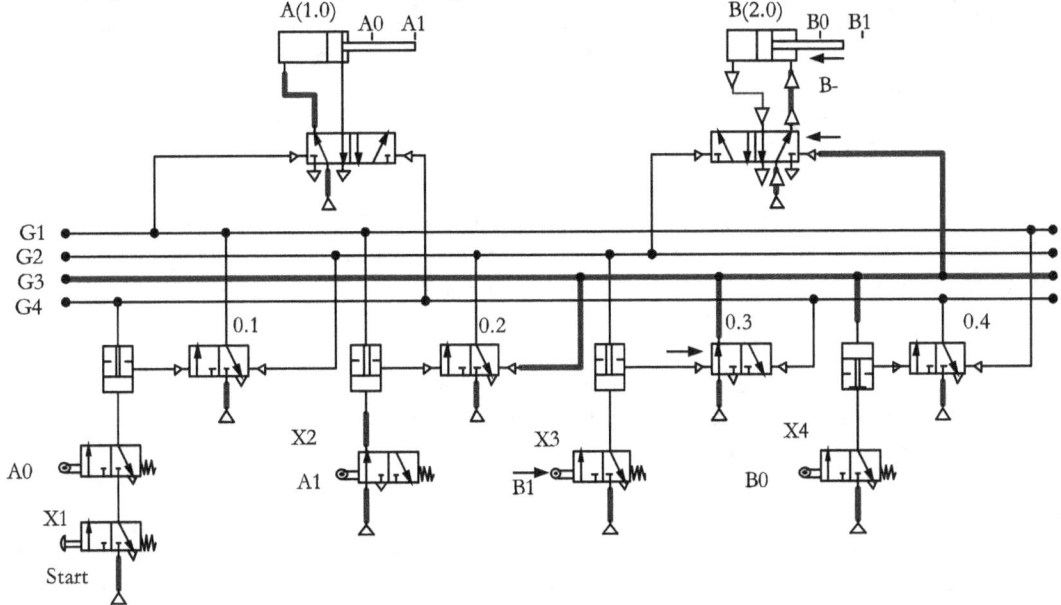

Figure 4.4(d) | The position of the circuit when x3 signal is applied

When Sensor B0 is Actuated

When cylinder B retracts, sensor B0 senses. As soon as the sensor B0 signal is applied, reversing valve 0.4 is set, and the supply reaches group G4, as shown in Figure 4.4(e). The valve 0.3 is reset, and cylinder A retracts (A-). Also, note that there is no signal conflict.

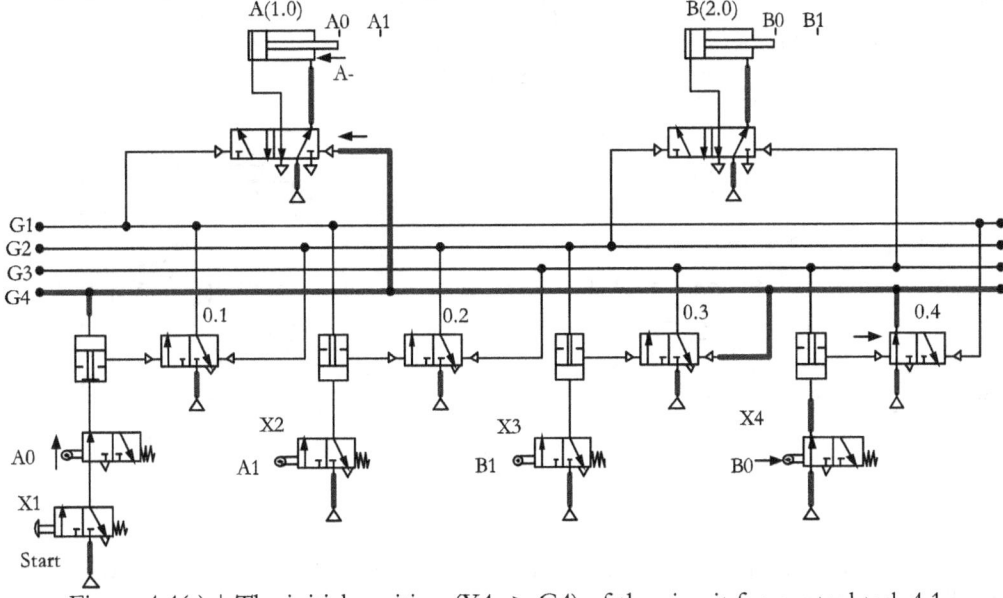

Figure 4.4(a) | The initial position (X4 -> G4) of the circuit for control task 4.1

Chapter 5 | Stepper Sequencer – Design and Operating Principles

Various logic functions are used in a pneumatic system to generate the necessary sequential output signals in response to system inputs, thereby controlling the actuators. Each step in the control sequence has a similar logical structure, except for the last stage; hence, it is advantageous to modularize the controller for each step. A module designed to meet the functional requirements of each step in the control sequence is usually called a step module. As is usual in a pneumatic circuit, the last stage in the sequential circuit is set initially with all other stages reset. Therefore, the design of the module for the last stage is slightly different from the design of the modules for other stages.

A step module consists of an integrated circuit with the following three logic functions: memory, AND, and OR. This module generates an output signal to execute a movement in the cycle after receiving signals from the process and the previous module. At the same time, it resets the preceding module via an OR valve.

Various manufacturers of pneumatic components offer step modules to meet the functional requirements of each step in sequential circuits. However, the technical implementation of the step module logic varies by manufacturer.

Step Modules

The basic step module, as shown in Figure 5.1(a), consists of a 3/2-way (or 5/2-way) memory valve, an AND element, and an OR element. The power supply port is P.

The step module is initialized by applying a signal to the port marked 'Initialization'. The memory valve is set with a signal from the previous stage in the sequence applied at port Y_n. The signal from the previous stage represents the previous group signal (memory valve output) logically combined in series with the AND element, which is actuated by the sensor connected at input X of the previous stage. The memory valve's output is taken as the output (A) of the step module. It is also used to reset the memory valve of the previous stage through port Z_n. The memory valve of the succeeding stage is set with a signal from the step module applied at port Y_{n+1}. The memory valve of the present stage is reset by the output of the succeeding stage's memory valve through port Z_{n+1}.

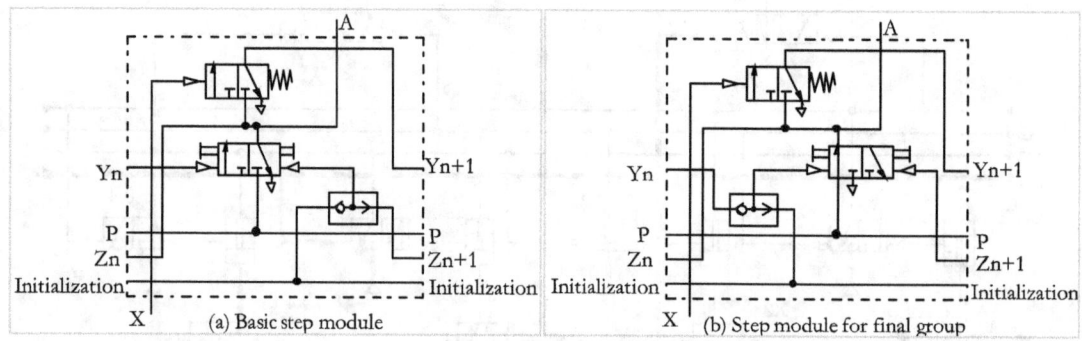

Figure 5.1 | Step modules

In the step module for the last group, the OR element is connected to set the step module's memory valve, as shown in Figure 5.1(b).

Stepper Sequencer

The step module is the fundamental element of the sequencer, controlling a step in the cycle. The sequencer must consist of as many modules as there are steps in the cycle, connected sequentially to complete the cycle. The modular sequencer provides control to each step of the machine operation and receives the corresponding feedback signal. The feedback signal initiates the next step in the cycle, and so on.

Circuit Design for a Stamping System (Two Groups)

Control Task 5.1 | Stamping System

The stamping operation uses a job-positioning cylinder A (1.0) [1A] and a stamping cylinder B (2.0) [2A]. Job positioning cylinder A (1.0) extends, bringing the job (work-piece) under stamping cylinder B (2.0). The cylinder B, then, extends and stamps the job. Cylinder A can return only after cylinder B has retracted fully. A pneumatic control circuit must be developed to implement this control task using step modules.

Solution

The notational form of representation of the control task is given in Figure 5.2.

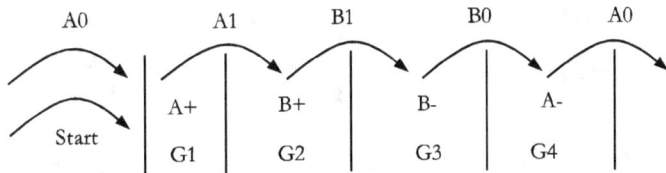

Figure 5.2 | The notational form of representation of the problem in control task 4.1

Control Circuit Using Step Modules

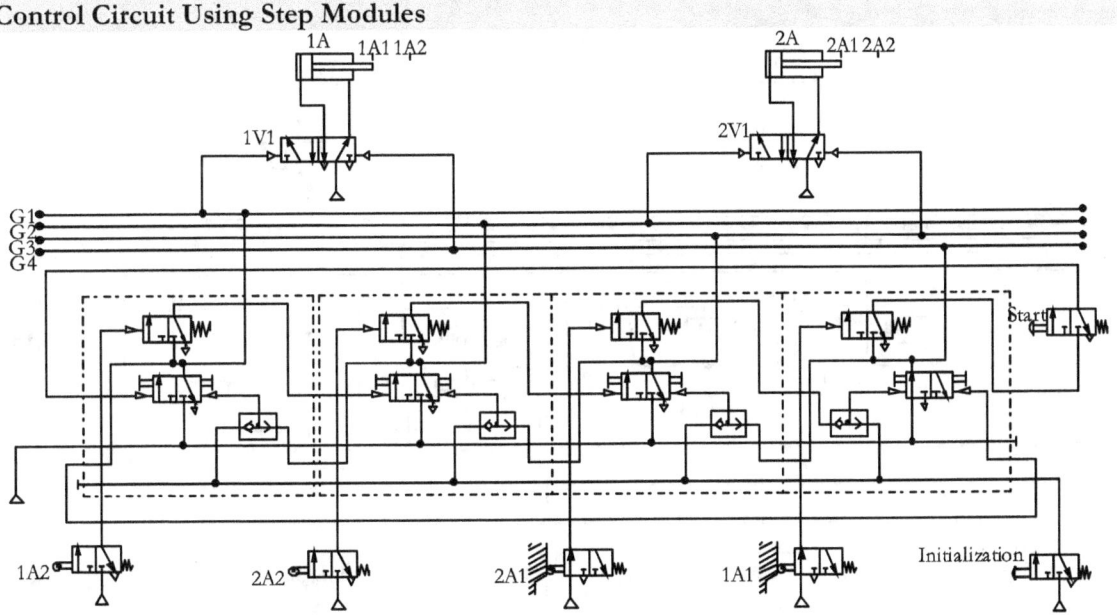

Figure 5.3 | A circuit for control task 5.1

A circuit for control task 5.1 is given in Figure 5.3. Various positions of the circuit are given in Figure 5.4(a), (b), (c), and (d).

The Position of the Circuit When the Start Pushbutton is Pressed (A+)

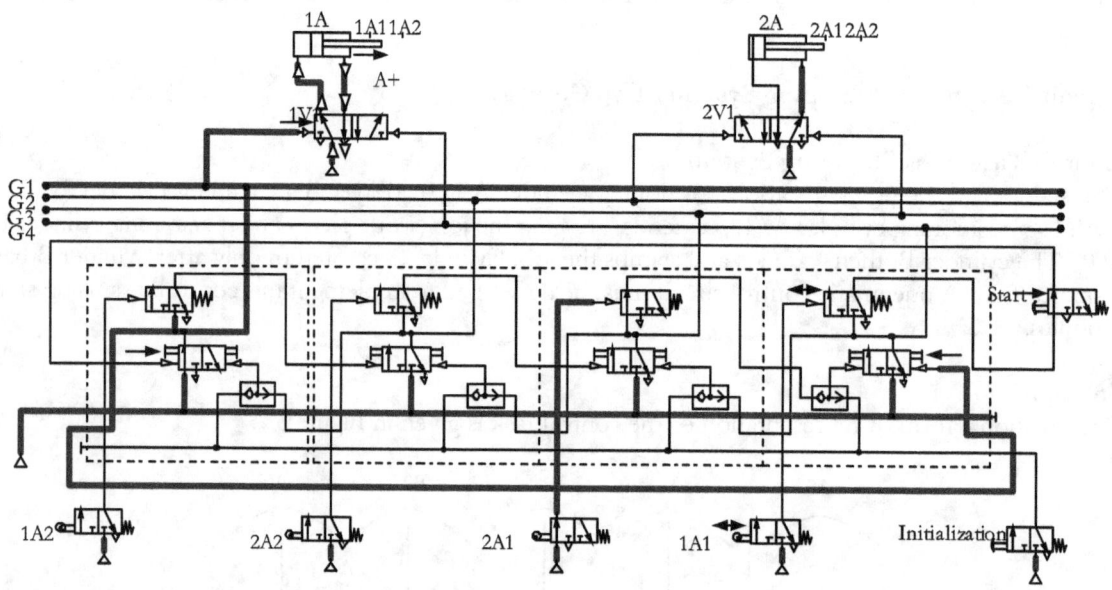

Figure 5.4(a) | The position of the circuit when the start pushbutton is pressed (A+)

The Position of the Circuit when Sensor 1A2 is Pressed (B+)

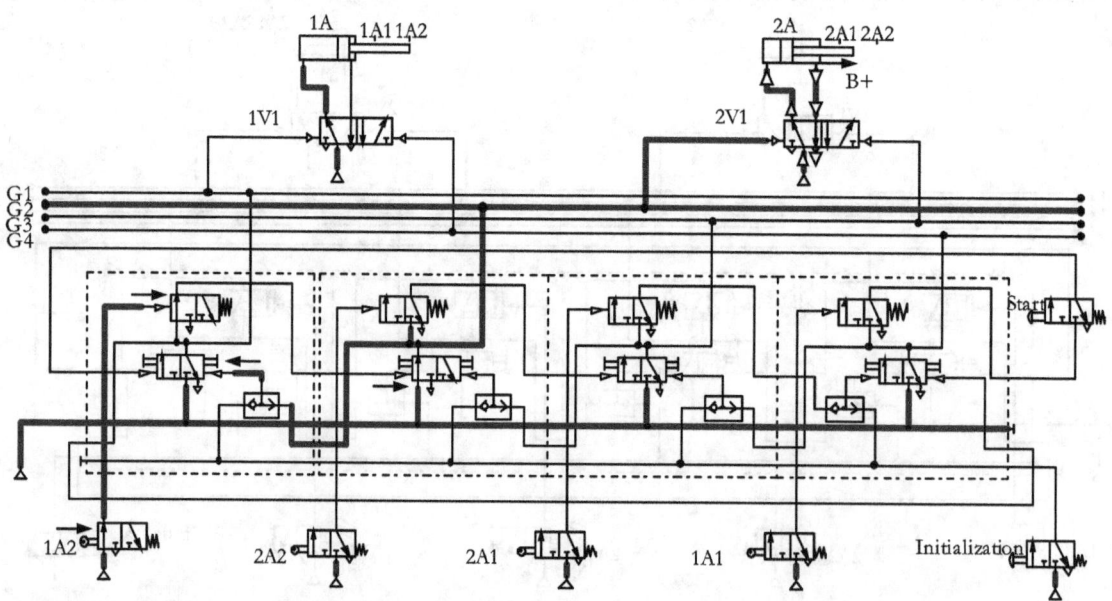

Figure 5.4(b) | The position of the circuit when sensor 1A2 is pressed (B+)

The Position of the Circuit When Sensor 2A2 is Pressed (B-)

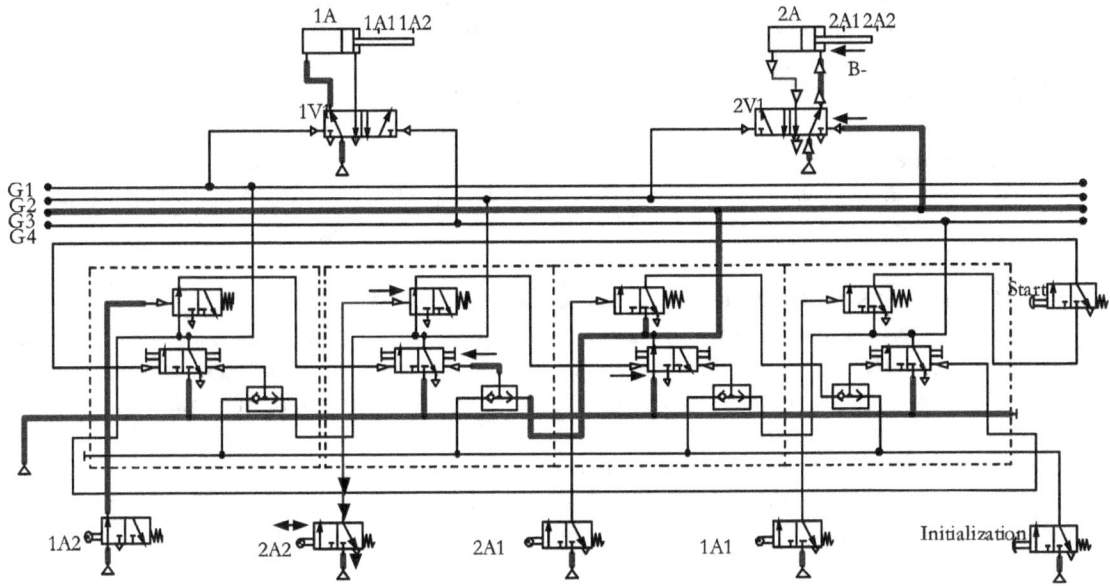

Figure 5.4(c) | The position of the circuit when sensor 2A2 is pressed (B-)

The Position of the Circuit When Sensor 2A1 is Pressed (A-)

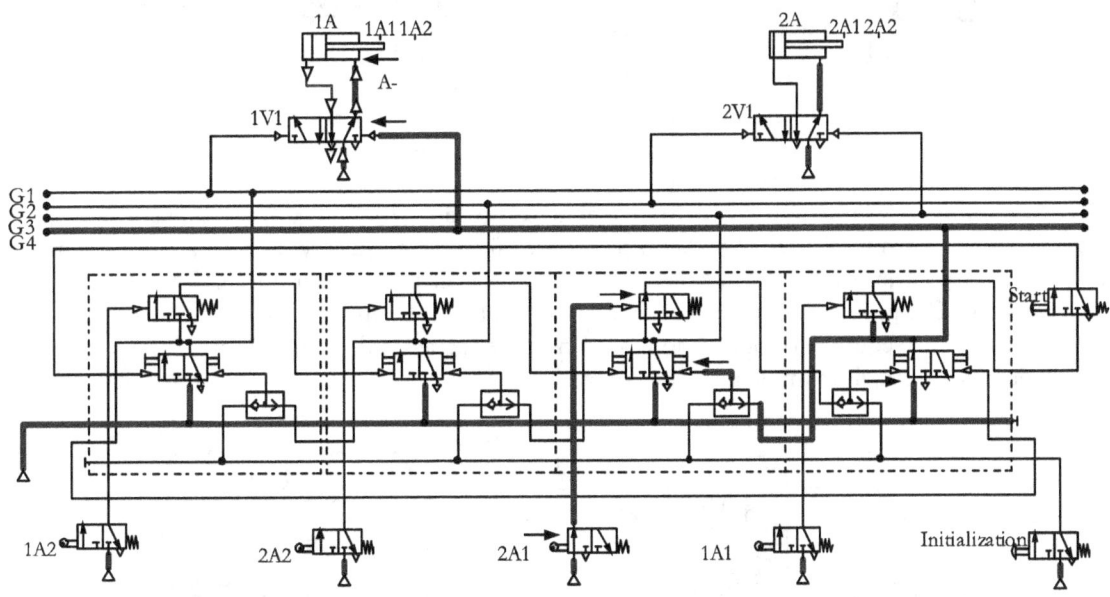

Figure 5.4(d) | The position of the circuit when sensor 2A1 is pressed (A-)

6 | Review Questions

1. What is signal conflict in multiple-actuator pneumatic circuits?
2. What is the effect of signal conflict in multiple-actuator pneumatic circuits?
3. What are the many ways to eliminate signal conflicts?
4. Explain the step-displacement diagram with a suitable sketch.
5. List a few disadvantages of using idle-return rollers for overcoming signal conflicts.
6. Briefly explain the principle of the cascade method as used for eliminating signal conflicts.
7. Draw a group changing cascade circuit for: (1) two groups, (2) three groups, and (3) four groups.
8. Briefly explain the principle of the shift register as used for eliminating signal conflicts.

7 | Additional Exercises for Circuit Development

7.1 | Pneumatically-actuated Lathe: The arrangement for an automatically operated pneumatic lathe is shown in Figure 7.1(a). First, cylinder 1.0 shifts the entire feed magazine to the transfer station. Then, cylinder 2.0 pushes the work-piece into the lathe chuck. Cylinder 1.0 and 2.0 then retract. The lathe now begins to run. Cylinder 3.0 moves the slide rest forward and back. Cylinder 4.0 then ejects the workpiece after the lathe chuck is released. The sequence of operations of these cylinders is shown in the displacement-step diagram [Figure 7.1(b)]. Develop a pneumatic control circuit to implement the given control task: (i) for one cycle, and (ii) for a continuous sequence of operations.

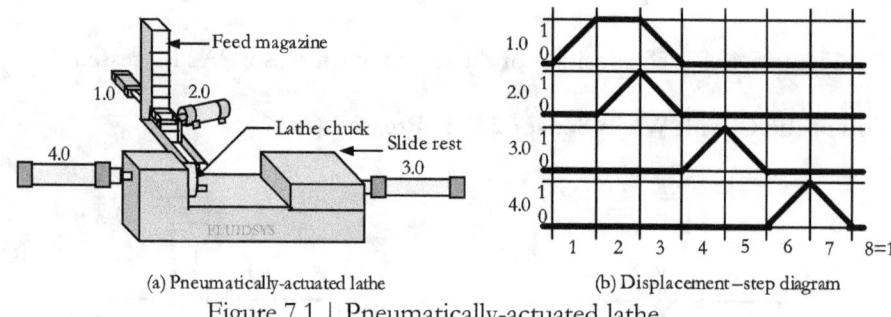

(a) Pneumatically-actuated lathe (b) Displacement–step diagram

Figure 7.1 | Pneumatically-actuated lathe

7.2 | Processing of Work-pieces: An arrangement, for 'feeding, clamping, processing and ejecting' of work-pieces, is shown in Figure 7.2(a). Workpieces are taken from the gravity-fed magazine and placed into the device by cylinder 1.0. Next, cylinder 2.0 extends to clamp the workpiece. Cylinder 3.0 is used for processing. The processing time (t) is adjustable via a timer. Cylinder 4.0 is used to eject the work-piece. The required sequence of operation is shown in the sequence diagram [Figure 7.2(b)]. Develop a pneumatic control circuit to implement the control task.

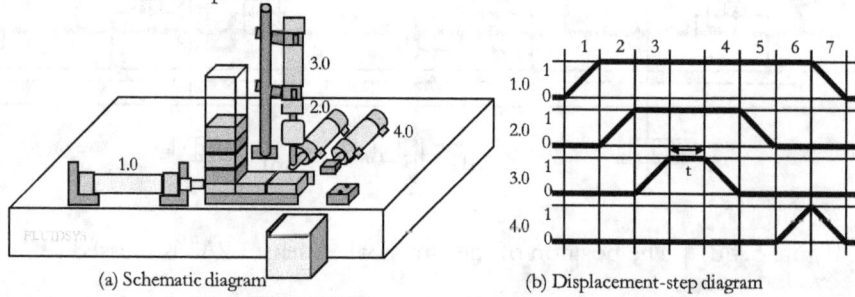

(a) Schematic diagram (b) Displacement-step diagram

Figure 7.2 | The processing of work-pieces

Note: Solutions to additional exercises are provided on the corresponding page numbers. 45-48

7.3 | Pneumatic Flanging Device: A pipe is to be flanged in two steps in a pneumatic flanging device, as shown in Figure 7.3(a). Cylinder 2.0 is in the extended position initially to act as a stop for the advancing pipe. Cylinder 1.0 extends and clamps the pipe, and subsequently, cylinder 2.0 retracts. Flanging cylinder 3.0 extends for the first time to perform the preliminary flanging operation, then retracts after a set time delay of t sec. The tool switchover is performed by cylinder 4.0, and flanging cylinder 3.0 is in action for a second time for the final flanging operation. The complete retraction of cylinder 3.0, after the set time delay of t sec, triggers the return motion of clamping cylinder 1.0 and tool switching cylinder 4.0, and the forward motion of position cylinder 2.0. The required sequence of operation is shown in the sequence diagram [Figure 7.3(b)]. Develop a pneumatic control circuit for the semi-automatic operation to implement the control task.

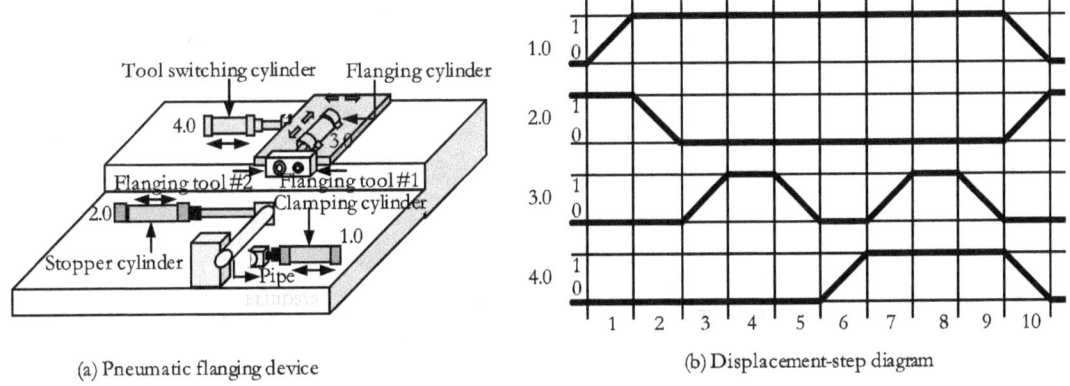

(a) Pneumatic flanging device   (b) Displacement-step diagram

Figure 7.3 | A pneumatic flanging device

7.4 | Pneumatic System for Reaming of Drill Holes: A pneumatic system is designed for reaming of drill holes, as shown in Figure 7.4(a). Cylinder 1.0 clamps the work-piece. Feed cylinder 2.0 moves fully forward and then backward, with the ability to differentially control speed. The feed cycle is repeated two more times. The required sequence of operation is shown in the sequence diagram [Figure 7.4(b)]. Develop a pneumatic control circuit to implement the control task.

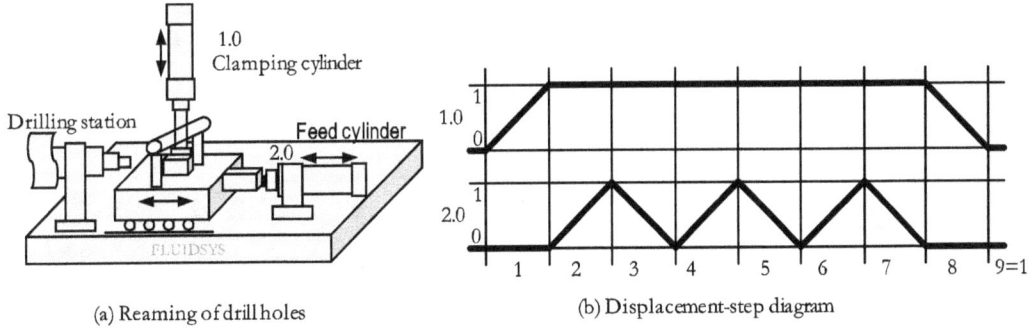

(a) Reaming of drill holes   (b) Displacement-step diagram

Figure 7.4 | A pneumatic system for reaming of drill holes

Appendix 1

Graphic Symbols for Pneumatic Components

Supply Elements

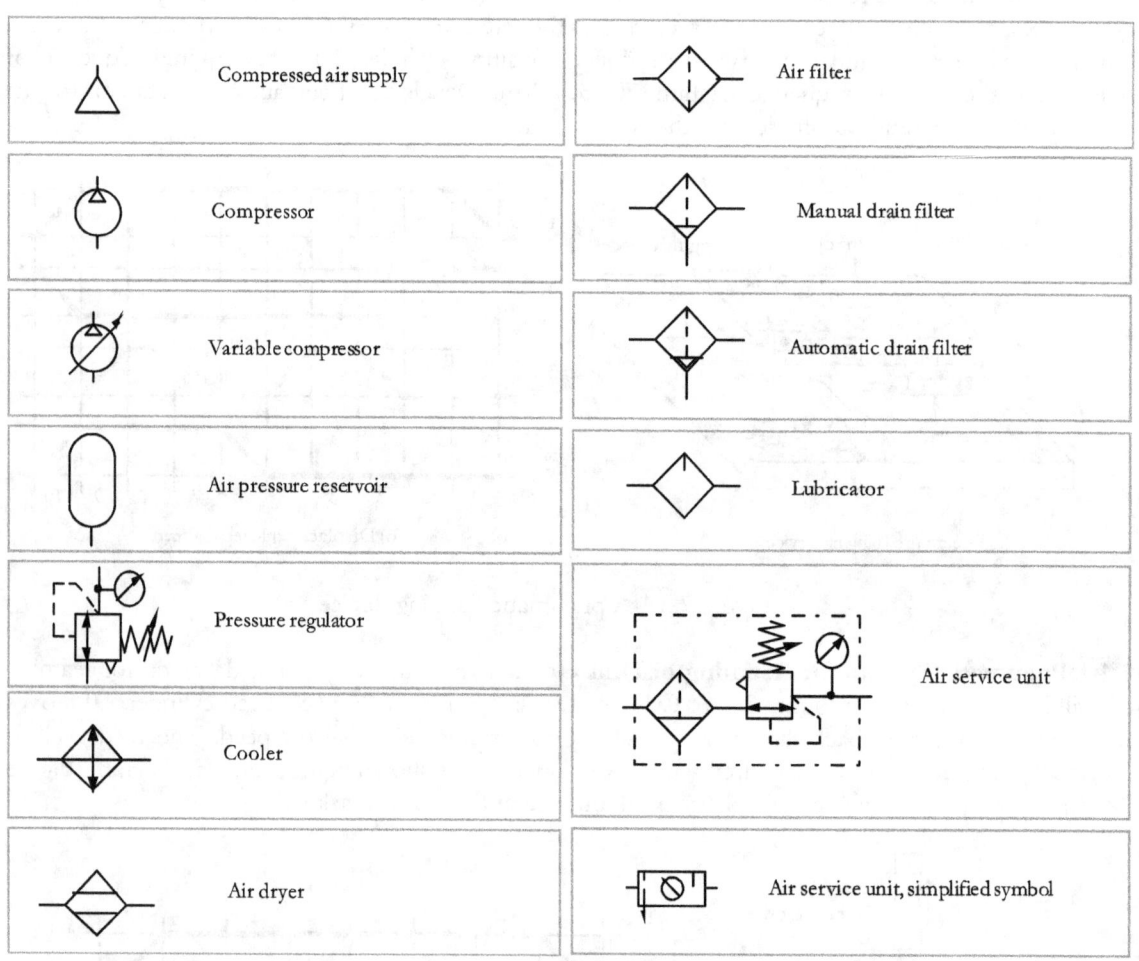

Actuators

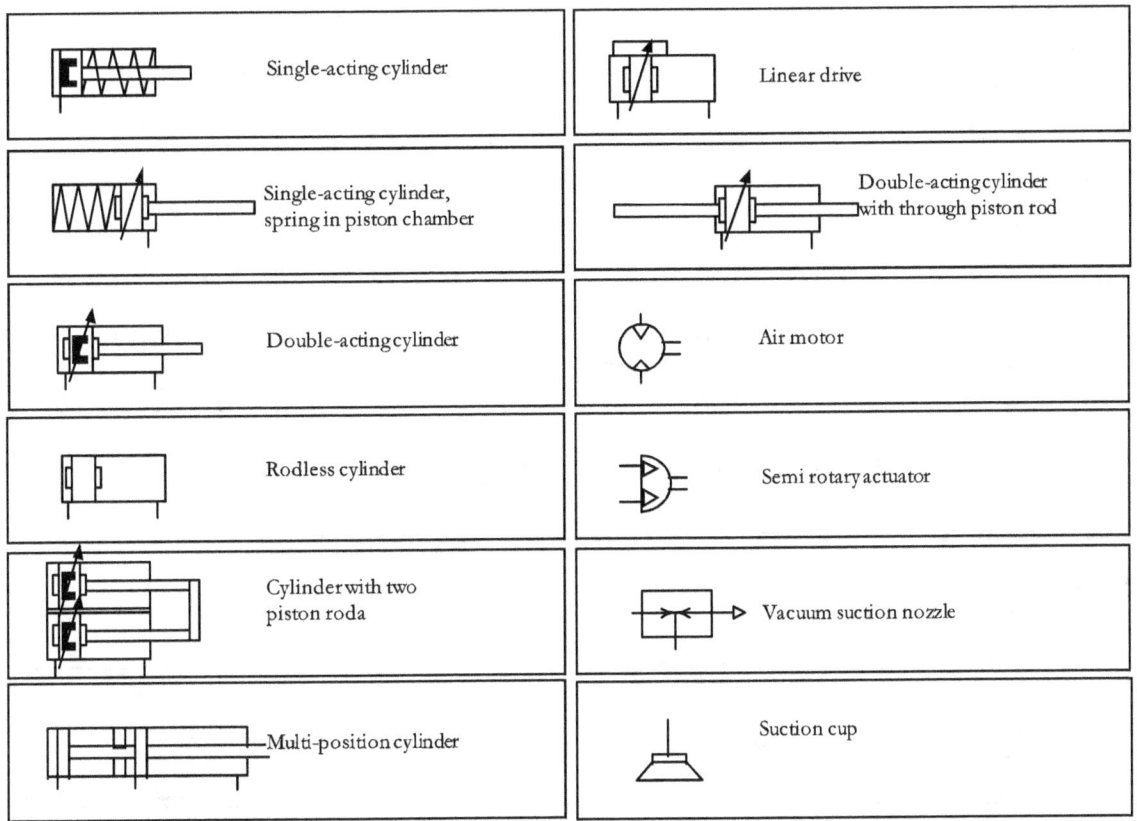

Shut-off Valves and Flow Control Valves

	2/2-way valve (NC) type		Check valve
	2/2-way valve (NO) type		Spring-loaded check valve
	3/2-way valve (NC) type		Orifice valve, adjustable type
	3/2-way valve (NO) type		Shuttle valve
	4/2-way valve		Two-pressure valve
	5/2-way valve		Quick-exhaust valve
	5/3-way valve, all closed centre position		Throttle valve with constant restriction
	5/3-way valve, open exhaust centre position		Throttle valve with adjustable flow control
	5/3-way valve, open pressure centre position		Throttle check valve, adjustable

Energy Transmission

Symbol	Name	Symbol	Name
——	Working line	—⊢	Energy tapping point closed
——	Control line	⌀	Pressure gauge
⌣	Flexible line	⊗	Pressure indicator
⊢ ⊣	Line connection, rigid	⌀	Flow meter
┼ ⌿	Line crossover	→⊢⊣←	Quick release coupling uncoupled without check valve
Exhaust symbol	Exhaust without pipe connection	→⊢⊣←	Quick release coupling connected without check valve
Exhaust symbol	Exhaust with pipe connection	—▷⊢⊣◁—	Quick release coupling uncoupled with check valve
Silencer symbol	Silencer	—▷⊢⊣◁—	Quick release coupling connected with check valve

Valve Groups

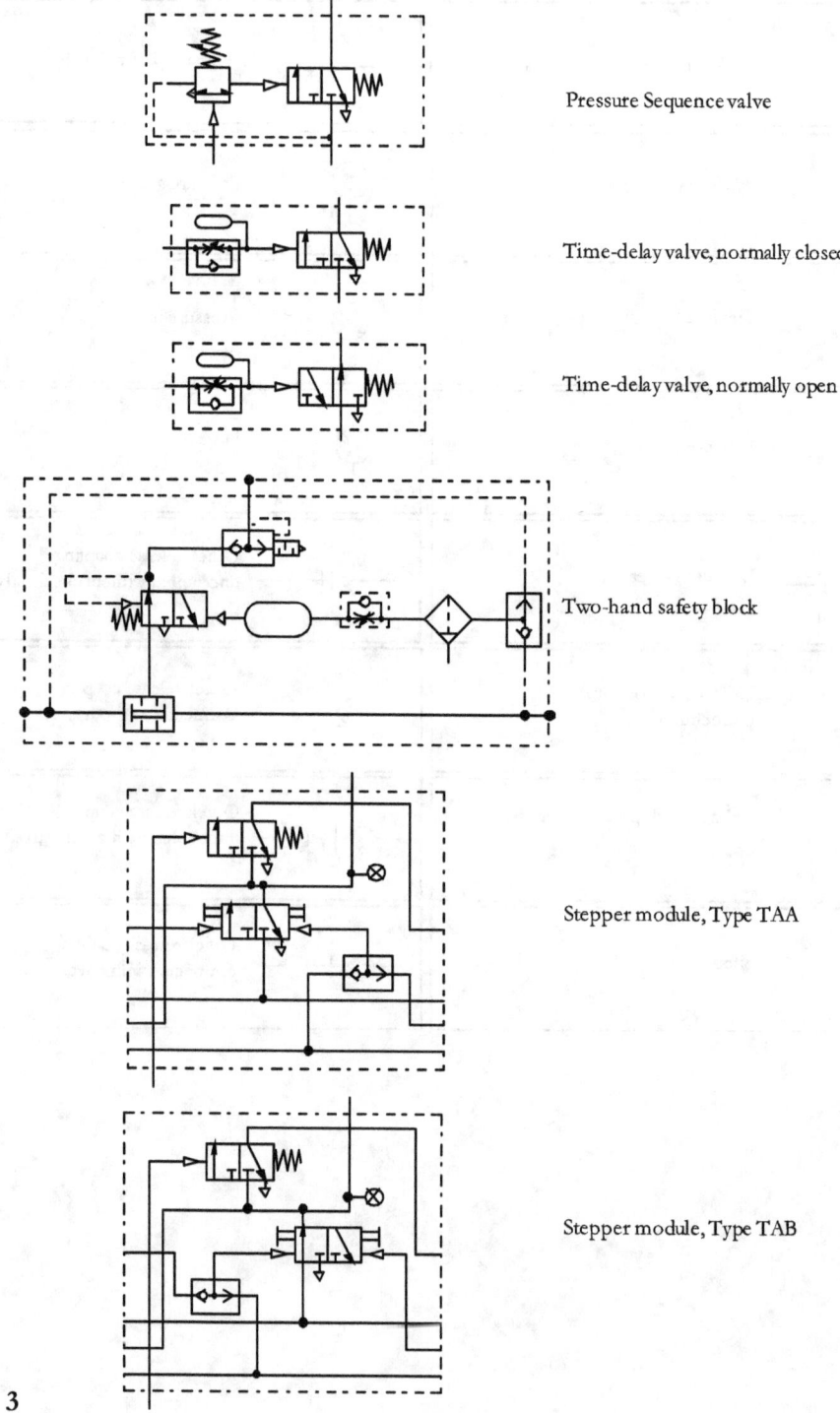

Pressure Sequence valve

Time-delay valve, normally closed

Time-delay valve, normally open

Two-hand safety block

Stepper module, Type TAA

Stepper module, Type TAB

3

Appendix 2

Designation of Pneumatic System Components Using Numbers

The components in the pneumatic circuit can be divided into groups, such as the energy supply group and actuator (working) groups. A pneumatic circuit generally consists of many actuators and control components. An actuator, along with all the associated control components, forms a working group.

The designations of pneumatic system components using numbers are given in Table A2.1. The general structure for the designation of a component in a pneumatic circuit is X.Y, where X represents the designation, such as 0, 1, 2, 3, etc., indicating the group to which it belongs. Y represents the designation (e.g., 0, 1, 2, 3, 4, 5) indicating the type of component within a group, such as an actuator, final control element, or processing and signal valves.

Table A2.1 | Designation of pneumatic system components using numbers

Element/Designation	X	.	Y	Example
Energy supply	0	.	1, 2, 3	0.1, 0.2, 0.3 ....
Working groups (One working group per actuator)				
Actuator			0	1.0, 2.0, 3.0 ......
Final Control Element			1	1.1, 2.1, 3.1 ......
Valves controlling forward motion	1, 2, 3...	.	(2, 4, 6...)	1.2, 2.4, (even numbers for Y)
Valves controlling return motion			(3, 5, 7...)	1.3, 2.5, (odd numbers for Y)
Speed control valve, forward motion			02	1.02, 2.02 ......
Speed control valve, return motion			01	1.01, 2.01 ......

Designation of Pneumatic System Components Using Letters

Alternatively, pneumatic components can be designated using letters. Table A2.2 gives the letter designations for pneumatic components.

Table: A2.2 | Designation of pneumatic system components using letters

Element	Designations
Energy supply elements	0Z1, 0Z2...
Actuators	1A, 2A...
Control elements	1V1, 1V2...
Signal elements for the return stroke of cylinders	1S1, 1S2, 2S1...

Representation of Valves Actuated in the Initial Position

A valve that is actuated in the initial position must be indicated in a circuit diagram with a trip cam shown alongside. A 3/2-DC normally-closed type roller valve, which is actuated in the initial position, is drawn in the circuit diagram in a manner, as shown in Figure A2.1.

Figure A2.1 | Representation of valves actuated in the initial position

Please refer to the information given in the textbook "Pneumatic Systems and Circuits - Basic Level (in the SI units)' and 'Pneumatic Systems and Circuits - Basic Level (in the English units)' by the same author.

Solutions for Additional Exercises

Solution for the Additional Exercise 7.1 (Pneumatically-actuated Lathe)

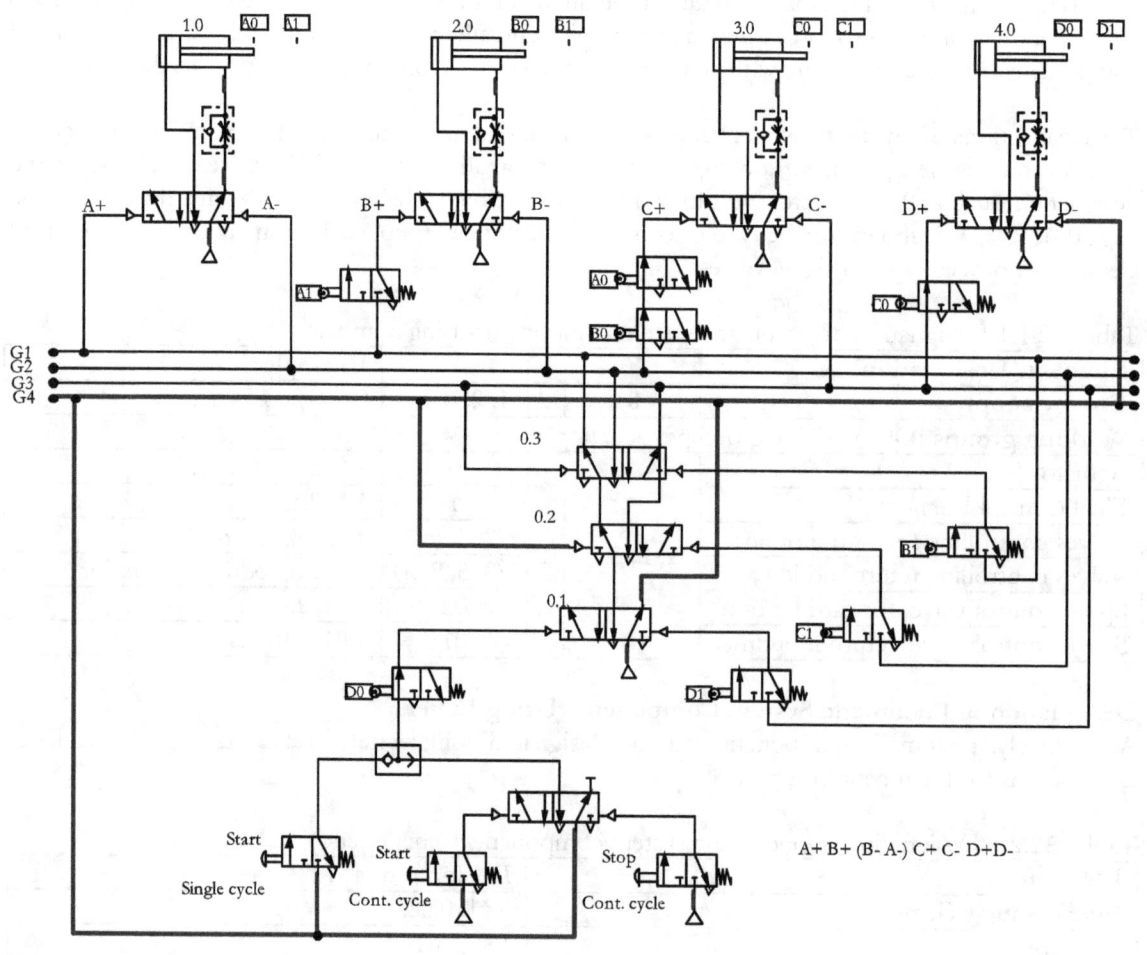

A+ B+ (B- A-) C+ C- D+ D-

Solution for the Additional Exercise 7.2 (Processing of Work-pieces)

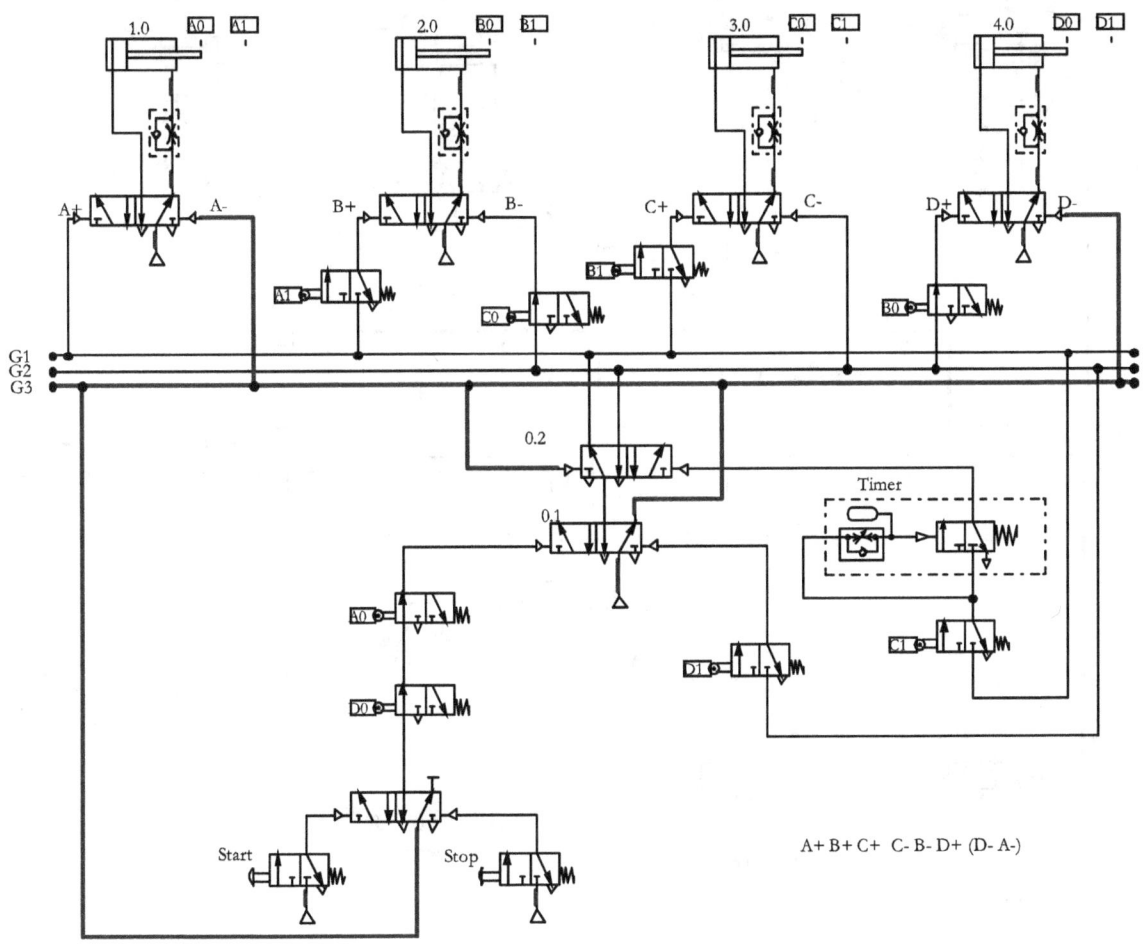

A+ B+ C+ C- B- D+ (D- A-)

Solution for the Additional Exercise 7.3 (Pneumatic Flanging Device)

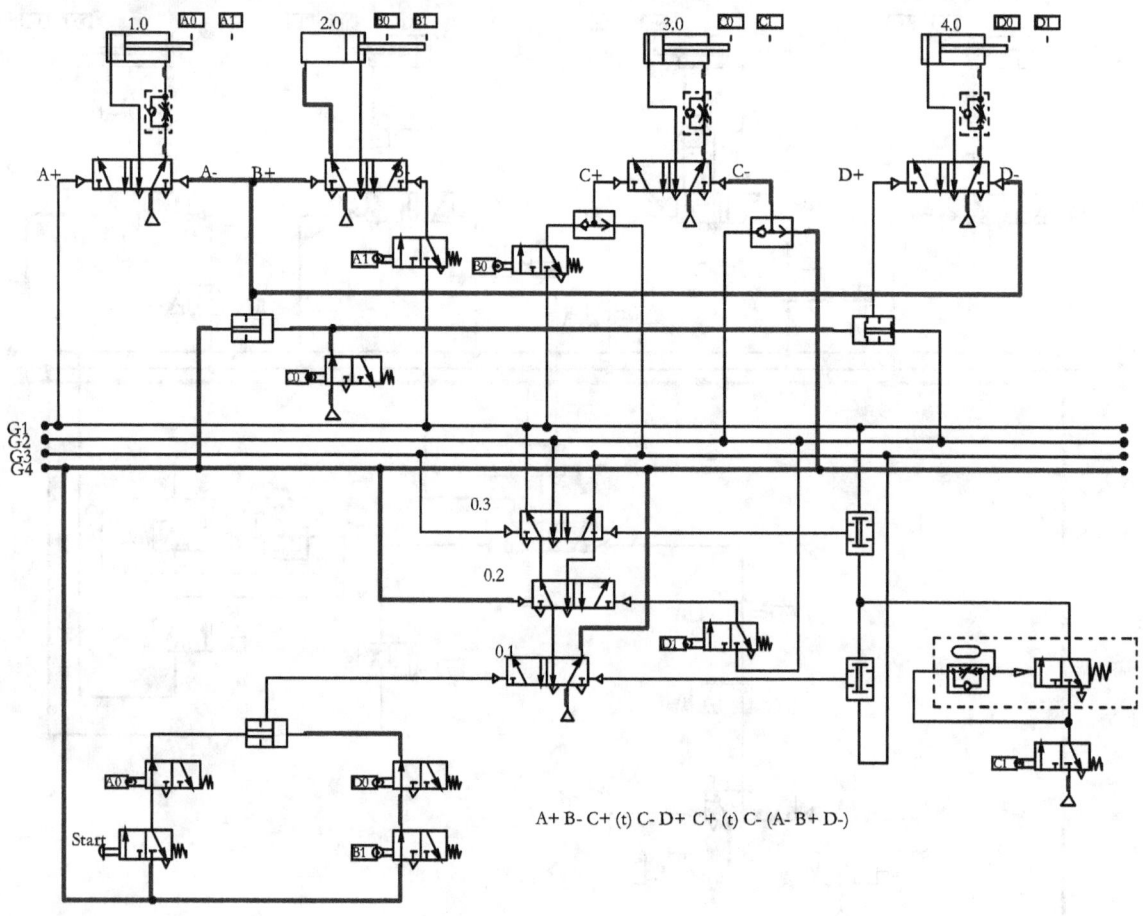

A+ B- C+ (t) C- D+ C+ (t) C- (A- B+ D-)

Solution for the Additional Exercise 7.4 (Pneumatic System for Reaming of Drill Holes)

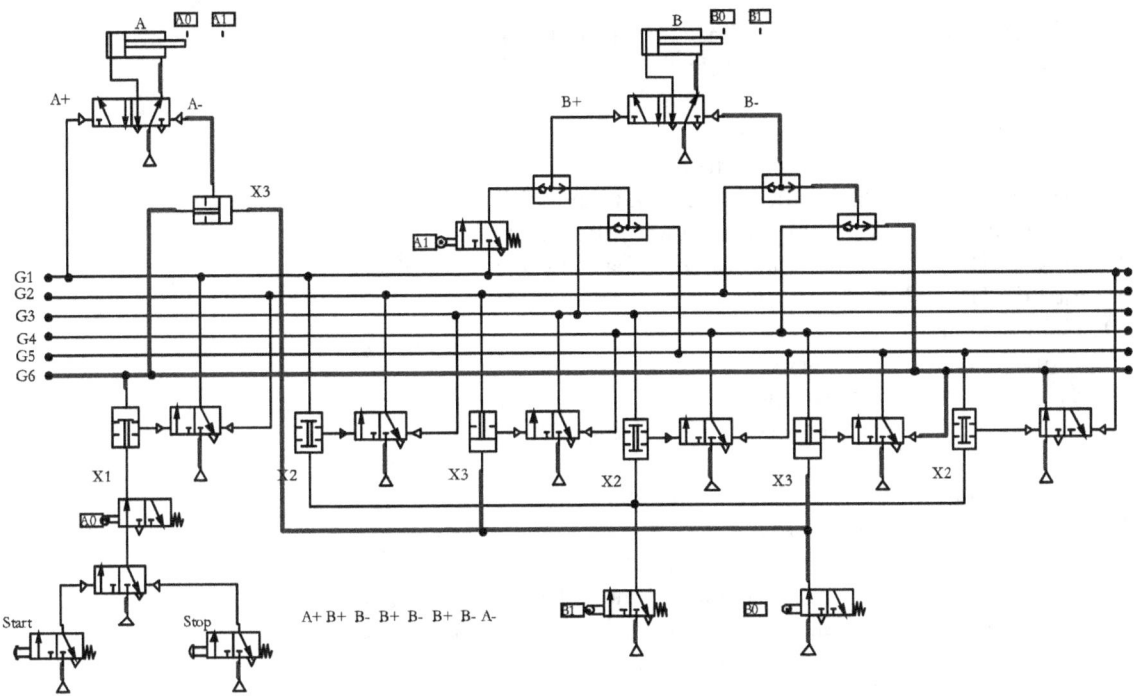

Fluid Power Educational Series Books

1. Pneumatic Systems and Circuits -Basic Level (In the SI Units)
2. Industrial Pneumatics -Basic Level (In the English Units)
3. Pneumatic Systems and Circuits -Advanced Level
4. Electro-Pneumatics and Automation
5. Design of Pneumatic Systems (In the SI Units)
6. Design Concepts in Pneumatic Systems (In the English Units)
7. Maintenance, Troubleshooting, and Safety in Pneumatic Systems
8. Industrial Hydraulic Systems and Circuits -Basic Level (In the SI Units)
9. Industrial Hydraulics -Basic Level (In the English Units)
10. Hydraulic Fluids
11. Hydraulic Filters: Construction, Installation Locations, and Specifications
12. Hydraulic Power Packs (In the SI Units)
13. Power Packs in Hydraulic Systems (In the English Units)
14. Hydraulic Cylinders (In the SI Units)
15. Hydraulic Linear Actuators (In the English Units)
16. Hydraulic Motors (In the SI Units)
17. Hydraulic Rotary Actuators (In the English Units)
18. Hydraulic Accumulators and Circuits (In the SI Units)
19. Accumulators in Hydraulic Systems (In the English Units)
20. Hydraulic Pipes, Tubes, and Hoses (In the SI Units)
21. Pipes, Tubes, and Hoses in Hydraulic Systems (In the English Units)
22. Design of Industrial Hydraulic Systems (In the SI Units)
23. Design Concepts in Industrial Hydraulic Systems (In the English Units)
24. Maintenance, Troubleshooting, and Safety in Hydraulic Systems
25. Hydrostatic Transmissions (HSTs) (In the SI Units)
26. Concepts of Hydrostatic Transmissions (In the English Units)
27. Load Sensing Hydraulic Systems (In the SI Units)
28. Concepts of Load Sensing Hydraulic Systems (In the English Units)
29. Electro-hydraulic Proportional Valves
30. Electro-hydraulic Servo Valves
31. Cartridge Valves
32. Electro-hydraulic Systems and Relay Circuits
33. Practical Book: Pneumatics - Basic Level
34. Practical Book: Electro-pneumatics - Basic Level
35. Practical Book: Industrial Hydraulics – Basic Level
36. Programmable Logic Controllers and Programming Concepts
37. Compressed Air Dryers
38. Hydraulic Circuits – Identification of Components and Analysis

For more details, please visit: **https://jojibooks.com**.

www.ingramcontent.com/pod-product-compliance
Lightning Source LLC
Chambersburg PA
CBHW080530220526
45465CB00006B/2658